INTRODUCTION TO INSECT PEST MANAGEMENT
Robert L. Metcalf and William H. Luckman, Editors

OUR ACOUSTIC ENVIRONMENT
Frederick A. White

ENVIRONMENTAL DATA HANDLING
George B. Heaslip

THE MEASUREMENT OF AIRBORNE PARTICLES
Richard D. Cadle

ANALYSIS OF AIR POLLUTANTS
Peter O. Warner

D1373914

THE MEASUREMENT OF AIRBORNE PARTICLES

THE MEASUREMENT OF AIRBORNE PARTICLES

RICHARD D. CADLE, Ph. D.

NATIONAL CENTER FOR ATMOSPHERIC RESEARCH
BOULDER, COLORADO

A WILEY-INTERSCIENCE PUBLICATION

JOHN WILEY & SONS

NEW YORK · LONDON · SYDNEY · TORONTO

Copyright © 1975 by John Wiley & Sons, Inc.

Library of Congress Cataloging in Publication Data:
Cadle, Richard D
  The measurement of airborne particles.

  (Environmental science and technology)
  "A Wiley-Interscience publication."
  Includes bibliographical references.
  1. Aerosols—Measurement. I. Title.

TD884.5.C3   628.5′3   75-22121
ISBN 0-471-12910-0

Printed in the United States of America

10 9 8 7 6 5 4 3 2 1

SERIES PREFACE
Environmental Science and Technology

The Environmental Science and Technology Series of Monographs, Textbooks, and Advances is devoted to the study of the quality of the environment and to the technology of its conservation. Environmental science therefore relates to the chemical, physical, and biological changes in the environment through contamination or modification, to the physical nature and biological behavior of air, water, soil, food, and waste as they are affected by man's agricultural, industrial, and social activities, and to the application of science and technology to the control and improvement of environmental quality.

The deterioration of environmental quality, which began when man first collected into villages and utilized fire, has existed as a serious problem since the industrial revolution. In the last half of the twentieth century, under the ever-increasing impacts of exponentially increasing population and of industrializing society, environmental contamination of air, water, soil, and food has become a threat to the continued existence of many plant and animal communities of the ecosystem and may ultimately threaten the very survival of the human race.

It seems clear that if we are to preserve for future generations some semblance of the biological order of the world of the past and hope to improve on the deteriorating standards of urban public health, environmental science and technology must quickly come to play a dominant role in designing our social and industrial structure for tomorrow. Scientifically rigorous criteria of environmental quality must be developed. Based in part on these criteria, realistic standards must be established and our technological progress must be tailored to meet them. It is obvious that civilization will continue to require increasing amounts of fuel, transportation, industrial chemicals, fertilizers, pesticides, and countless other products and that it will continue to produce waste products of all descriptions. What is urgently needed is a total systems approach to modern civilization through which the pooled talents of scientists and engineers, in cooperation with social scientists and the medical profession, can be focused on the development of order and equilibrium to the presently disparate segments of the human environment. Most of the skills and tools that are needed are already in existence. Surely a technology that has created such manifold environmental problems is also capable of solving them. It is our hope that this Series in Environmental Sciences and Technology will not only serve to make this challenge more explicit to the established professional but that it also will help to stimulate the student toward the career opportunities in this vital area.

Robert L. Metcalf
James N. Pitts, Jr.
Werner Stumm

PREFACE

The importance of airborne particles to humanity must be obvious to most persons; it hardly needs to be emphasized here. For example, we all live in a great aerosol system, our atmosphere, and the particles it contains have vast influences on our climate and on our health. Much of our technology involves the production, use, and often the elimination of airborne particles. Growing concern about man's impact on the quality of our environment, and specifically about the increasing burden of atmospheric particles for which man is responsible on local, regional, and even global scales, has accelerated investigations of airborne particles by colleges, universities, government agencies, and industrial laboratories. Measurement of the size distributions, shapes, and concentrations of the particles is essential to such investigations; the techniques and equipment for making such measurements is the subject of this book.

The proposition that the development and expansion of technology will occur to fill a need is axiomatic, and the art and science of aerosol particle measurement is no exception. In 1955 I wrote a small book entitled *Particle Size Determination* which was the first monograph devoted to this subject. Particles in various media such as powders, pastes, suspensions in liquids, and suspensions in air were considered. Since then a number of excellent books on particle measurement have been written that have enabled the reader to keep up to date in this rapidly expanding field. Now both the theory and art of airborne particle measurement is so extensive that the treatment of the subject in depth nearly requires the publication of a monograph dealing with that subject alone. I have undertaken the task of writing such a book because I have directed and carried out laboratory and field investigations of aerosols ever since being asked to participate in a study of Los Angeles smog many years ago. Some of the field investigations have become adventures when they involved determining the contributions to atmospheric trace constituents of the rain forests of Panama and Amazonia, volcanoes in various parts of the world, and nuclear explosions.

This book has been written primarily as an aid to the scientist or engineer faced with the problem of making aerosol particle measurements. However, it should also be useful to a person desiring to learn more about this field, and as a textbook or source of supplementary material for college or university courses concerned with aerosol science and technology. The book deals to a greater extent than most monographs with commercially developed and marketed instruments. I believe that this is essential, since many of the greatest improvements made in recent years in the particle measurement art have occurred in industrial laboratories. Furthermore, reference to specific manufacturers should aid the reader in locating needed equipment. Clearly, not all commercial sources of airborne particle measuring equipment could be included, and the omission of any such source should not be interpreted to be derogatory. Furthermore, the selection of illustrations of commercial equipment in most cases depended on their being supplied by the manufacturer.

The first chapter deals to a considerable extent with particle statistics and data presentation, emphasizing recent developments. The organization of the chapters dealing with actual measurement techniques is fairly straightforward, each chapter discussing first theory and then methodology. However, the organization of the material on impaction required some soul searching on my part, because impaction techniques can be used both to collect and to classify particles according to size. Impaction techniques designed merely to collect particles are described in Chapter 2, and multistage impactors are discussed in Chapter 5. I hope that this arrangement is not too confusing. The last chapter deals with the important subject of isokinetic sampling.

More space has been devoted to filtration and to optical microscopy than to many of the other subjects. This is partially because filtration and optical microscopy are very old and very useful techniques that have received a great deal of attention. Furthermore, they are available to most scientists and engineers with the outlay of little or no money. Another reason for discussing optical microscopy in great detail is that although microscopes are part of the equipment of a great many laboratories, few persons know how to use them to best advantage.

I am pleased to acknowledge the help of many persons in the preparation of this volume. I especially wish to thank Ms. Nadine Perkey who coped so successfully with my nearly illegible handwriting while typing the manuscript, Dr. J. P. Lodge, Jr., for his helpful scientific and technical comments, and Ms. Kay Redman who helped polish the text.

R. D. CADLE

Boulder, Colorado
January, 1975

CONTENTS

THE MEASUREMENT OF AIRBORNE PARTICLES

1

INTRODUCTION

Aerosols can be defined as any relatively stable suspension of particles in a gas, especially in air, including both the continuous (gaseous) and discontinuous (particulate) phases. Since World War II, especially during the last decade, great advances have been made in our knowledge of the behavior of aerosols and in the techniques available for studying them. Perhaps the greatest incentive for the present accelerated research on aerosols has been the increasing concern about the impact of man on his environment, particularly the extent to which he is polluting the atmosphere—locally, regionally, and worldwide. Another incentive for this research is the need in many modern devices for closer tolerances that can be achieved only by manufacturing them in relatively dust-free rooms, the so-called clean rooms. Studies of airborne particles are an important aspect of industrial hygiene, weather modification, and studies of the natural atmosphere. Furthermore, investigations of the production, characterization, and behavior of aerosols has become a scholarly discipline.

Often, the most important part of any aerosol investigation is measuring the suspended particles. Here, measurement refers to the determination of size distributions, concentrations, and shapes. Such measurements, of course, constitute one aspect of fine-particle technology, and methods for making them have been described in this context in many publications.[1-4] During recent years, a proliferation of techniques, especially automatic methods, has developed for measuring the particles in various environments (powders, liquid suspensions, and aerosols, for example); and the measurement of airborne particles is now a specialty. In fact, many of the methods developed for the measurement of aerosol particles are applicable only to particles suspended in a gas.

1. SOME DEFINITIONS

1.1. Particles

The term "particle" can be defined as any object having precise physical boundaries in all directions, and the meaning of the term "individual particles" in an aerosol is often obvious. However, when the suspended particles are aggregates, when there are particles of different types such as of different chemical composition, or when the size distributions are very wide or otherwise complex, the investigator must be careful to precisely define the particles with which he is concerned. Then the methods selected for making the desired measurement must be consistent with that definition. If the suspension contains aggregates, the investigator must decide whether he should measure the aggregates, the individual particles in the aggregates, or both. A common error may occur when an aerosol contains two or more types of particles, each type having a greatly different size distribution. If the investigator is interested in only one of the particle types and is not aware of the presence of the others, gross measurement errors may result.

Errors from this source can arise, for example, when an aerosol is prepared with an aspirator by dispersing a dilute suspension of particles in a volatile liquid. Many of the droplets produced contain the insoluble particles, and evaporation of the volatile liquid leaves the particles in suspension in air. However, if the liquid contained traces of dissolved, nonvolatile material, such as a wetting agent, evaporation of the droplets will produce an aerosol containing, in addition to the desired particles, large numbers of very small particles consisting only of the nonvolatile material. Even if only one type of particle is in suspension, if the size distribution is very wide, a decision often must be made with regard to the size range of interest and the appropriate measurement technique to be employed.

1.2. Particle Size

Particle size and particle shape also need precise definition if they are to be measured. The need to define particle shape if it is to be described quantitatively is usually apparent, but the similar need to define particle size is often overlooked. One of the most common and frustrating omissions in the scientific and technical literature dealing with fine particles is the failure to state whether particle "size" refers to radii or diameters. But, even if this is indicated, the term diameter (or radius) is unambiguous only for spheres. When all the particles in a system such as an aerosol have the same shape (such as cubic), then the problem of defining diameter in terms of the dimensions of the particles is relatively easy. However, when the particles vary markedly in shape, definition is much more difficult.

Diameters of irregular particles can be defined in terms of the geometry of the individual particles or in terms of their physical properties. Diameters of the former type are often described as statistical because they have much meaning only for averages of the measurements of a large number of irregular particles. Sieving as a method of size determination (which is not discussed in this book because it is seldom if ever appropriate for aerosol particles) can be considered to yield statistical diameters, since the results depend directly on the dimensions of the particles and are averaged by the method for a large number of particles.

Diameters defined in terms of physical properties are determined by measuring properties such as light scattering, sedimentation rates, or rates of Brownian diffusion.

Perhaps the most widely used statistical diameter is the diameter of a circle whose area is the same as that of the area of the particle projected onto a surface. It was originally used primarily for particle measurement by optical microscopy, but it is being used more and more in connection with automatic and semiautomatic sizing methods.

This definition is easy to apply when the particles are regular in shape, but it is almost impossible to apply to certain aggregates. An example is the particles shown in the electron micrograph of Figure 1-1. They were collected with impactors mounted on U.S. Air Force RB-57F aircraft from the stratosphere at about 18 km altitude a few weeks after extentensive brush fires in California, and they may have been particles of wood smoke.[5] One way to treat such aggregates is to determine the sizes of the individual ("ultimate") particles and the number of such particles in each aggregate measured.

Another difficulty often arises when airborne droplets are collected on hard surfaces, for example, by impaction. The droplets tend to spread on the surface and may, upon pulling together, leave satellite droplets. The resulting pattern even may provide a means for identifying the major constituent of the droplets.[6] Figure 1-2 is an electron micrograph of particles collected from the Antarctic atmosphere near McMurdo Sound;[7]

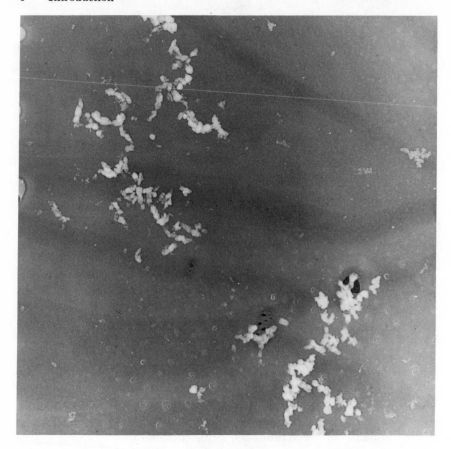

Figure 1-1. Particles collected by impaction from the midlatitude stratosphere at 18 km altitude on 21 October 1970. The distance across the electron micrograph is 6 μm. From Cadle, R. D., EOS, Transactions, American Geophysical Union, **53**, 812 (1972). Copyright by the American Geophysical Union. Magnification 19,314×.

Figure 1-3 is an electron micrograph of particles collected from the fume from the lava fountains of Kilauea volcano in Hawaii in 1967.[8] The patterns are typical of sulfuric acid droplets, a common atmospheric constituent. If size distributions are to be determined from measurements of the images of the collected droplets, attempts must be made to estimate the original droplet volume. Some methods for doing this are discussed in the section on microscopy.

Other definitions of diameter are based on the assumption that the length (l), breadth, (b), and thickness (t) of the particle can be measured.

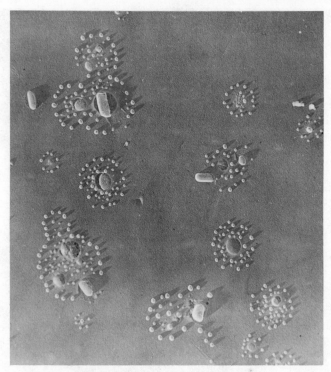

Figure 1-2. Electron micrograph of particles impacted from Antarctic air.[7] Field of view is 6 μm across. The particles were probably impure sulfuric acid.

These can be presented separately or the diameter can be defined as the arithmetical average of the three dimensions

$$d = \left(\frac{1}{3}\right)(l + b + t) \qquad (1\text{--}1)$$

The dimensions also can be used to define the diameter as the length of the side of a cube of equal volume

$$l = \sqrt[3]{lbt} \qquad (1\text{-}2)$$

Martin[9] proposed the following definition of diameter. It is the distance between opposite sides of the particle, measured crosswise of the particle, and on a line bisecting the projected area (Figure 1-4). The diameters must always be measured in the same direction if the results are to be statistically significant. For example, if the measurements are

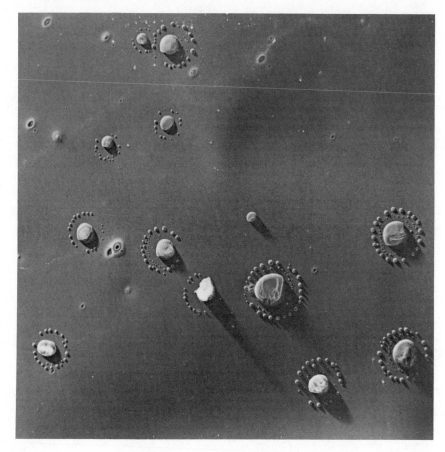

Figure 1-3. Electron micrograph of particles collected by impaction from the fume from the lava fountains of Halemaumau crater of Kilauea volcano during the eruption of 1967–1968. The particles consisted of impure sulfuric acid. Distance across micrograph is 25 μm[8].

made with a microscope, the direction parallel to the bottom of the field is convenient. Tomkeiff[10] and Moran[11] have suggested that the relationship between Martin's diameter and the area per unit volume (specific surface) of a powder can be calculated from the expression

$$\text{Martin's diameter} = \frac{4}{D_p S_v}$$

where D_p is the packing density and S_v is the specific surface.

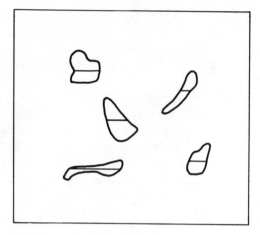

Figure 1-4.   Martin's diameters.

Feret's diameter[12] is the distance between two tangents on opposite sides of the particle. As in the case of Martin's diameter, the tangents must be parallel to an arbitrarily fixed direction.

Elongated particles such as fibers may be described merely by length, by thickness, or by both.[13]

1.3.   Particle Shape

Defining particle shape is often even more difficult than defining particle size.

Two general approaches to determining and reporting the shapes of particles are based on (1) measuring the dimensions of individual particles, and (2) estimating the mean geometric properties from two different properties of the aerosol, which are influenced to different extents by the shape of the particles. The first approach has already been implied; that is, measuring the length, breadth, and thickness of a large number of the collected particles. Heywood[14] has discussed this approach to determining particle shape and the effect of shape on the results of various types of particle size determinations. He proposed that the particle be assumed to be resting on a plane in the position of greatest stability. The breadth, b, he defined as the distance between two parallel lines tangent to the projection of the particle on the plane and placed so that the distance between them is as small as possible. The length, l, is the distance between parallel lines tangent to the projection and perpendicular to the lines defining the breadth. The thickness, t, is the distance between two planes parallel to the plane of greatest stability and tangent to the surface of the particle. Heywood defined flakiness, F, as b/t and elongation, E, as l/b.

Hausner[15] has proposed a method for characterizing particle shape based on the dimensions of a rectangle drawn around the projection of the particle. He proposed the use of three dimensionless ratios, called the elongation factor, the bulkiness factor, and the surface factor to characterize the shape of the particle regardless of its size. This, of course, is a two-dimensional treatment, as are several definitions of diameter described above but, to be precise, the definition of shape and diameter is a two-dimensional treatment, as are several definitions of diameter plate- or flake-type particles. Hausner suggested that the rectangle of minimum area drawn around the particle projection be used. The elongation of the particle is given by

$$x = a'/b' \qquad (1\text{-}3)$$

where x is the elongation factor, and a' and b' are the side lengths of the rectangle, a' being the longer. Hausner also defines a bulkiness factor

$$y = \frac{A}{a'xb'} \qquad (1\text{-}4)$$

where y is the bulkiness factor and A is the projected area of the particle. Thus, y is unity for a particle that just fills the rectangle but is usually much less than this.

Since, for particles of identical shapes, the ratio of surface area to volume decreases with increasing particle size, the rectangle dimensions can-

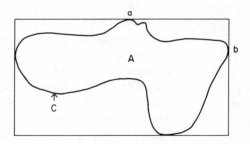

Figure 1-5. Characterization of particle shape using the method of Hausner;[15] a and b are the lengths of the sides of the rectangle, A is the projected particle area and C is the circumference.

not be used to characterize particle surface. For a sphere, the cross-sectional area A is $d^2\pi/4$ and the circumference C is $d\pi$, so $C^2 = 4\pi A$. Hausner defines a dimensionless surface factor, z, which is the ratio of the square of the circumference of the projected area A of the particle to the square of the circumference of the sphere of cross-sectional area A:

$$z = C^2/4\pi A \tag{1-5}$$

The more the particle deviates from spherical shape, the larger will be the value of z. Other workers have used very similar approaches. For example, Wadell[16] has defined sphericity as the ratio of the surface area of a sphere to that of a particle of equal volume.

The approach to estimating particle shape based on measuring average properties of particles has been discussed in great detail by Herdan[17] and involves the use of the so-called shape factors. He points out three functions of such factors:

1. They provide a means for comparing particle size determination results obtained by different methods.

2. They can be used to convert the results of size determinations in terms of "equivalent spheres."

3. They can transform mean diameter squared and cubed into mean particle surface and mean particle volume, respectively.

Surface and volume shape factors are defined as $\alpha_s = S/d^2$ and $\alpha_v = V/d^3$, respectively, where S is the surface area, V is the volume, and d the diameter of the particle. The values of α_s and α_v will, of course, depend on the methods used for obtaining S, V, and d. For example, the "projected" diameter, d_p, can be defined, as mentioned earlier, to be that of a circle having the same area as the projected image of the particle when viewed in the direction perpendicular to the plane of greatest stability. Similarly, the "equivalent" diameter, d_e, is the diameter of a sphere having the same volume as the particle. Shape factors are generally used in connection with averages for a large number of particles. Thus, shape factors may be defined by the equations

$$\bar{V} = a_v \, \bar{d}_p{}^3 \tag{1-6}$$

and

$$\bar{S} = a_s \, \bar{d}_p{}^2 \tag{1-7}$$

where the bars refer to arithmetic means, defined later.

The average volume of a particle in terms of the equivalent diameter is

$$\bar{V} = (1/6) \, \pi \bar{d}_e{}^3 = \alpha_v \, \bar{d}_p{}^3 \tag{1-8}$$

therefore,

$$\bar{d}_e = \bar{d}_p \sqrt[3]{\frac{6\alpha_v}{\pi}} \tag{1-9}$$

We can also compare $\bar{d}_p$ with the mean surface diameter, $\bar{d}_s$ obtained, for example, by sedimentation:

$$\alpha_s \bar{d}_p^{\,2} = \bar{d}_s^{\,2} \pi \tag{1-10}$$

and

$$\bar{d}_s = \bar{d}_p \sqrt{\frac{\alpha_s}{\pi}} \tag{1-11}$$

The value of α_s will, of course, depend upon the method used for determining the mean surface diameter.

Heywood[14] has differentiated between the proportions of particles and their geometrical shape as they influence the values of the shape factors. He derived the formula

$$\alpha_0 = \alpha_v F E^{1/2} \tag{1-12}$$

where F and E are the flakiness and elongation as defined by Heywood and described earlier, and α_0 is the shape factor for an equidimensional particle. He then deduced an equation for α_s in terms of α_v and α_0:

$$\alpha_s = 1.57 + C \left(\frac{\alpha_0}{F}\right)^{4/3} \left(\frac{E+1}{E}\right) = 1.57 + C\alpha_v \left(\frac{E+1}{E^{1/3}}\right) \tag{1-13}$$

C is a constant which must be determined experimentally.

We can generalize the principle of the use of shape factors by stating that the ratios between diameters obtained by different particle-sizing methods are related to average particle shapes and are essentially constant for a given type of aerosol. The proportionality factors are the shape factors. These are usually defined as the proportionality constants between mean diameters determined microscopically for individual particles and those determined for the entire aerosol, as by sedimentation or light scattering.

In spite of the rather elaborate methods described above for indicating particle shape, it is helpful to demonstrate the shapes by photomicrographs and electron micrographs. Scanning electron micrographs are especially useful for studying the irregularities of particle surfaces because of the great detail they provide (Figure 1-6). Stereo pairs are also often useful to indicate shape in three dimensions. An example is the stereo pair of photomicrographs of sodium chloride dendrites shown in Figure 1-7. These deposited while an aerosol of sodium chloride particles was

Figure 1-6. Scanning electron micrograph of microfossils. Courtesy of Applied Research Laboratories. Magnification 10,000×.

being prepared by condensation of sodium chloride vapor. Optical devices can, of course, be obtained for observing stereo pairs. However, with a little practice, anyone with reasonably normal vision can learn to view such pairs in three dimensions without the aid of special optical devices. The viewer looks between the two members of the pair and with his eyes brings the two images together to form a three-dimensional image. The viewer also sees the uncombined members of the pair as peripheral images.

1.4. Concentrations

Concentrations of particles in an aerosol can be defined in terms of particle number, volume, or mass, and in terms of gas volume or mass. The most important aspect when presenting concentration data is to insure that sufficient information is given that the interpretation can be quantitative. For example, if concentration is expressed in terms of particle number, but for a limited size range within a wide distribution, the limits of the size range must be specified. This is obvious, but is often ignored in the literature. If the concentrations are expressed in terms of

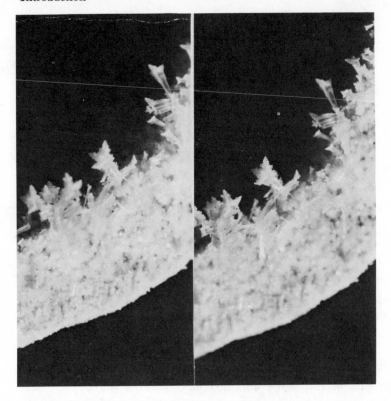

Figure 1-7.   Stereo pair of sodium chloride dendrites.

volume of gas, such as μg of particulate matter per cubic meter of air, the pressure of the aerosol should be indicated, especially if the concentration has been recalculated to some standard temperature and pressure, such as 273°K and 760 torr.

2.   PARTICLE STATISTICS

2.1.   Classification

The statistics of fine particles has been thoroughly reviewed in the excellent book by Herdan[17] and has been discussed to some extent in connection with particle sizing by other authors.[1-3] The present discussion emphasizes those aspects of particle statistics that are especially appropriate for aerosols.

Since suspensions of particles in gases never consist of particles that are exactly the same size (and the sizes often vary greatly) analysis of particle size data by statistical methods is often very enlightening. Such analysis can furnish information about both means of particle sizes and distributions of the sizes about the means.

When particle size determinations are made by optical or electron microscopy, the size of a large number of particles must be measured and the data arranged in such a way that the size distribution is apparent. The task of analyzing the data can be greatly simplified by dividing the entire size range of interest into a number of size intervals and tabulating the number of particles measured that fall into each group. This process is known as classification of the data; each interval is termed a class and the particle size limits are termed class boundaries. Ten to twenty classes are usually convenient since considerable information may be discarded with less than ten classes, and when more than twenty classes are used the process is unnecessarily tedious. Conventionally, the midpoint of each interval is called the class mark, d_i, where i is the number of the interval. The number of particles in each interval is called the frequency (f_i), and the total number of particles measured is denoted by n. Table 1-1 shows the results of such classification for a hypothetical determination of the diameters of 300 particles. Presumably, these particles would have been collected by some mechanism such as impaction from a known volume of an aerosol so that the concentration of the particles as well as the size distribution could be determined.

Table 1-1. Distribution of the Diameters of 300 Particles.

Class number (i)	Class boundaries (in μm)	Frequencies (f)	Class marks (d_i)
1	1.0– 2.0	5	1.5
2	2.0– 3.0	9	2.5
3	3.0– 4.0	11	3.5
4	4.0– 5.0	28	4.5
5	5.0– 6.0	58	5.5
6	6.0– 7.0	60	6.5
7	7.0– 8.0	54	7.5
8	8.0– 9.0	36	8.5
9	9.0–10.0	17	9.5
10	10.0–11.0	12	10.5
11	11.0–12.0	6	11.5
12	12.0–13.0	4	12.5

A size distribution such as that of Table 1-1 is obtained in the belief that the sample is representative of the entire population of interest. The results are then used to infer the size distribution for the entire population. The size distribution for airborne particles can specify the fraction of the aerosol assigned to the several classes in which case, as Fisher[18] has pointed out, the distribution may be (1) a limited number of fractions adding up to unity, (2) an infinite series of finite fractions adding up to unity, or (3) a mathematical expression representing the size distribution. However, when reporting size distributions of aerosol particles, especially of particles in the atmosphere, it is usually desirable to present the data in such a manner that the particle concentrations for various size intervals can be estimated.

The class intervals of classified data need not be identical but they should bear some simple relationship to one another. If the diameters are believed to be log-normally distributed, as described later, it is useful to make the class intervals equal differences between the logarithms of the diameters instead of equal differences between the diameters themselves. Logarithmic classification is especially useful for atmospheric particles, which always have a very wide size distribution. Furthermore, the frequencies can be for properties other than numbers of particles, especially total surface or weight in the class intervals.

Open-ended classification can also be used. The class for the smallest particles includes all particles smaller than the lower end of the last complete interval, or the class for the largest particles includes all particles larger than the upper end of the first complete interval, or both. Open-ended classification, although frustrating to the user of the data, often cannot be avoided because of the limitations of the measurement techniques.

2.2. The Arithmetic Mean

The arithmetic mean, or average, is the most commonly used statistical indicator of central tendency. It is commonly used in connection with aerosol particles, especially when the experimentally determined distributions are closed or the frequencies approach zero asymptotically at both ends, as in the case of the normal distribution. When they are open ended, as is practically always the case for experimentally determined distributions of atmospheric aerosols, the arithmetic mean is seldom very significant.

The arithmetic mean is one of a class of statistical values that may be defined in terms of "moments." The kth moment about the origin for classified data is defined by the equation:

$$m_k = \frac{1}{n} \sum_{i=1}^{h} d_i^k f_i \qquad (1\text{-}14)$$

where h *is* the number of class intervals. When the data are unclassified, all the f_i are equal to unity and $h = n$.

The first moment about the origin, m_1, is the arithmetic mean and will be indicated by a bar over the appropriate symbol. Thus, the arithmetic mean diameter for classified data is defined as

$$\bar{d} = \frac{1}{n} \sum_{i=1}^{h} d_i f_i \qquad (1\text{-}15)$$

and the average mass, $\bar{m}$, average surface, $\bar{s}$, and average volume, $\bar{v}$ may be similarly defined. For unclassified data equation (1-15) reduces to the familiar definition of arithmetic mean:

$$\bar{d} = \frac{\sum d}{n} \qquad (1\text{-}16)$$

The arithmetic mean can, of course, be calculated by means of equations (1-15) or (1-16), but the procedure is often tedious when there are large numbers of observations.

Methods for simplifying the calculation are described in many books on statistics, such as Hoel[19] and Wallis and Roberts.[20] One method depends on the selection of a class mark, d_0, near the center of the distribution and calculating values of a variable, u, defined by the equation

$$u_i = \frac{d_i - d_0}{c} \qquad (1\text{-}17)$$

where c is the class interval. Then equation (1-15) becomes

$$\bar{d} = \frac{1}{n} \sum_{i=1}^{h} (c u_i + d_0) f_i \qquad (1\text{-}18)$$

$$= c \frac{1}{n} \sum_{i=1}^{h} u_i f_i + d_0 \frac{1}{n} \sum_{i=1}^{h} f_i \qquad (1\text{-}19)$$

Also,

$$\bar{d} = c\bar{u} + d_0 \qquad (1\text{-}20)$$

Since the u's usually are smaller than the d's, the calculation is greatly simplified.

Table 1-2. Transformation for Calculating the Arithmetic Mean and Standard Deviation from the Data of Table 1-1.

$d_i(\mu m)$	f	u	uf	$u^2 f$
1.5	5	−5	−25	125
2.5	9	−4	−36	144
3.5	11	−3	−33	99
4.5	28	−2	−56	112
5.5	58	−1	−58	58
6.5	60	0	0	0
7.5	54	1	54	54
8.5	36	2	72	144
9.5	17	3	51	153
10.5	12	4	48	192
11.5	6	5	30	150
12.5	4	6	24	144
Totals	300		71	1,375

The hypothetical data of Table 1-1 have been recalculated using this type of transformation to produce Table 1-2. The class mark d_0 was chosen to be 6.5 μm. From equation (1-20) and Table 1-2 we obtain

$$\bar{d} = \frac{1}{300} (71) + 6.5 = 6.74$$

The method can also be used to calculate arithmetic means of surface and weight distributions by substituting surface or weight fractions of the total surface or weight for the frequencies.

2.3. Spread of the Size of the Particles

When a size distribution is closed-ended or the frequency is believed to approach zero asymptotically at both ends, the calculation of some number which represents the scatter about the means is often useful. For a truly closed-ended population of particles, and, of course, for any given set of measurements of size, which is always closed-ended, the range difference between the highest and lowest values can be indicated. The range is often reported for concentrations of atmospheric variable constituents, including particles, especially in connection with air pollution monitoring. Although the range is useful, it has the weakness that it depends to a considerable extent on the number of observations from which it is computed, tending to increase with increasing number of observations. It also suffers from the fact that one extraordinarily high or low value can tremendously increase the range. As Wallis and Roberts[20] point out, these weaknesses are especially dangerous when ranges are used for comparing different groups of observations.

A less-well-known measure of scatter is the mean (or average) deviation. This can be defined as the arithmetic mean of the absolute values of the deviations from the arithmetic mean of the dispersion. Absolute values, of course, must be used or they would tend to cancel one another.

There are at least two major disadvantages to the use of mean deviations.[20] One is that the mean deviations of several groups of measurements cannot be combined to produce the mean deviation of the combined group, as can be done with the standard deviation, which is discussed below. The second drawback is that the mean deviation is seldom useful for "statistical inference."

When the data have been classified, the mean deviation can be calculated from the expression $\Sigma \mid d_i - \bar{d} \mid f_i/n$. Even for unclassified data, there are short cuts. If we let A be the sum of the diameters greater than, and B the sum of those less than $\bar{d}$, and let a be the number of observations greater than and b the number less than $\bar{d}$, in both cases ignoring the values of d equal to $\bar{d}$, the mean deviation is given by $[A - B - (a - b)\bar{d}]/n$.

A theoretically better method for indicating the spread of the size distribution is the standard deviation, which is based on the concept of moment, as defined above. The second moment about the origin could be used as a measure of spread (scatter), but when we use the arithmetic mean as the measure of central tendency, it is advantageous to define moments about this mean as a measure or spread. The kth moment about the arithmetic mean for classified data is defined by the equation

$$m_k = \frac{1}{n} \sum_{i=1}^{h} (d_i - \bar{d})^k f_i \qquad (1\text{-}21)$$

The second moment about the mean is the variance, m_2, and can itself be used as a measure of variation. However, it is more convenient to have a measure of variation in the same units as the data, so the square root of the variance, called the standard deviation, s, is usually selected. Thus,

$$s = \frac{1}{n} \sqrt{\sum_{i=1}^{h} (d_i - \bar{d})^2 f_i} \qquad (1\text{-}22)$$

Unlike the standard deviation, the variance is not a good descriptive statistic but it can be used to compare populations and to combine samples.[17,20] Both m_2 and s are sometimes defined with n replaced by $n - 1$.

The standard deviation, like the arithmetic mean, can be computed directly from its definition. However, it can more easily be computed using the transformation of equation (1-17). It is readily shown that

$$s = c \sqrt{\frac{\Sigma u_i^2 f_i}{n} - \bar{u}^2} \tag{1-23}$$

For unclassified data, equation (1-22) becomes

$$s = \sqrt{\frac{\Sigma d_i^2}{n} - \bar{d}^2} \tag{1-24}$$

where for simplicity, the summation limits have been omitted.

Values of $u^2 f$ computed from Table 1-1 are given in Table 1-2. From these values and equation (1-23) we obtain:

$$s = \sqrt{\frac{1375}{300} - \left(\frac{71}{300}\right)^2} = 2.1 \ \mu m$$

The standard deviation often seems more meaningful when it is related to the magnitude of the mean. A common practice is to express the standard deviation as a percentage of the mean, $100 \ s/\bar{d}$. It is called the relative standard deviation or the coefficient of variation.

Clearly, the means of samples are only useful if they are nearly the same as the mean of the population from which the sample was taken. An estimate can be made of the closeness of the approximation by obtaining the standard deviations of the means of the samples about the population mean. Usually, the "true" value of the population mean will not be known, but the arithmetic mean of the sample average can be used. The standard deviation of the mean can be estimated from the equation

$$s_m = \frac{s}{\sqrt{n}} \tag{1-25}$$

where s is the standard deviation of the population. Since the latter value usually is unknown, the standard deviation of a sample is used. This operation, of course, introduces an error, but it is not serious for large samples. Equation (1-25) demonstrates that, in general, the accuracy of the estimation of the arithmetic mean of a population of particles increases with the square root of the number of particles measured.

The sizes of the particles in some aerosols have a normal distribution, as defined later. For such a distribution, the interval $(\bar{d} - s)$ to $(\bar{d} + s)$ includes about 68% of the measurements, and the interval $(\bar{d} - 2s)$ to $(\bar{d} + 2s)$ includes about 95% of the measurements. When the particle sizes are log-normally distributed, the standard deviations of the logarithms of the sizes tends to follow this rule.

Another class of indicators of scatter about the mean (or any indicator of central tendency) is the so-called quantiles. They are defined by position on the scale of cumulative frequency. The measure of central tendency is usually the median value, which can be defined as that particle size for which half of the particles are larger and half smaller, and which is itself a quantile. Many quantiles are used often enough to have special names such as quartiles, deciles, and centiles. For particles, they are usually numbered from the smallest to the largest diameters. Thus, one-fourth of the diameters are smaller than the first quartile, and the second quartile is the median.

2.4.   Skewness and Peakedness

The third moment about the mean is defined by equation (1-21) with $k = 3$. A zero third moment corresponds to symmetry about the arithmetic mean, whereas a positive third moment indicates a large right tail for a plot of f versus d, with d increasing to the right; such a distribution is said to be positively skewed, or skewed to the right. The third moment furnishes only a qualitative measure of skewness, since it depends in part on the units in which particle diameter is expressed. A dimensionless measure of skewness can be obtained by dividing the third moment about the mean by the cube of the standard deviation $(m_3/m_2^{3/2})$.

A measure of peakedness, also called kurtosis, is obtained by dividing the fourth moment about the mean by the fourth power of the standard deviation (m_4/m_2^2). It is usually larger for more peaked distributions, but there are often exceptions. The value of m_4/m_2^2 for normal distributions is 3.

2.5.   Definitions of Central Tendency

Two definitions of central tendency have already been presented, namely for the arithmetic mean and the median. Numerous other definitions have been developed for particle measurements that are used under various circumstances. One of these is the mode, which is the value of greatest frequency. Of course, a monotonically increasing or decreasing size distribution has no mode, and many distributions have more than one mode (bimodal, etc). The mode has the advantages that it is readily found from a plotted frequency distribution and that, when compared with the arithmetic mean, it provides an indication of the degree of skewness. However, it has questionable value in descriptive statistics.

Numerous means exist in addition to the arithmetic mean, and each is useful for certain types of comparison. Many of these means can be defined by the general equation for classified data

$$(\bar{d}_{qp})^{q-p} = \sum_{i=1}^{i=h} \frac{f_i}{n} d_i^q \Big/ \sum_{i=1}^{i=h} \frac{f_i}{n} d_i^p \qquad (1\text{-}26)$$

The definitions of the means are determined by the values of q and p. For example, the arithmetic mean is defined by assigning $p = 0$ and $q = 1$. Note that all of the means defined by equation (1-26) are, thus, defined in terms of powers of ratios of moments about the origin (equation 1-14) where q and p indicate the order (first, second, etc.) of the moment.

Different authors tend to use different names for the same definition of diameter, or conversely, different definitions for the same name. The nomenclature used as follows is fairly conventional but not universal.

The diameter of the particle whose surface is the arithmetic mean of the particle surfaces is the mean surface diameter and is defined by equation (1-26) when $q = 2$ and $p = 0$:

$$\bar{d}_{20} = \sqrt{\frac{1}{n} \Sigma f_i d_i^2} \qquad (1\text{-}27)$$

It is, thus, the square root of the second moment about the origin, and is especially useful when considering the surface behavior of particles.

The mean weight (or volume) is the diameter of the particle whose volume is the arithmetic mean of the volumes:

$$\bar{d}_{30} = \sqrt[3]{\frac{1}{n} \Sigma f_i d_i^3} \qquad (1\text{-}28)$$

This mean is especially useful when one is interested in the contributions of particles of various sizes to the total mass of particles.

Only in the trivial case where the particles are all the same size will these means be identical. It is not unusual to find that the arithmetic mean of the diameters of aerosol particles is in the neighborhood of 0.1 μm, but that the diameter of the particle of average mass is ten times as great. This is because a few very large particles may account for most of the mass or volume of a sample. In general, these averages are related to one another by the inequality $\bar{d}_{10} < \bar{d}_{20} < \bar{d}_{30}$. The magnitude of the differences among these diameters gives an indication of the variation in particle size of the sample.

The linear mean diameter is defined as $\bar{d}_{21}$; that is, $\Sigma\, f_i d_i{}^2/\Sigma\, f_i d_i$.

The surface mean diameter is defined as $\bar{d}_{32}$; that is $\Sigma\, f_i d_i{}^3/\Sigma\, f_i d_i{}^2$. Mugele and Evans[21] call it the Sauter diameter and describe a method for calculating the efficiency of "atomizing" a liquid from it. The surface mean diameter is often used in connection with surface area determinations. Herdan[17] restricts the above definition of the surface mean diameter (which he calls the volume-surface mean particle size) to spheres and cubes, and defines the mean in more general terms by the equation

$$\bar{d} = \Sigma d_i f_s(d_i^2)\, f_i/\Sigma f_s(d_i^2)\, f_i \tag{1-29}$$

where $f_s(d_i{}^2)$ is a function representing the surface of the particles. The reciprocal of $\bar{d}_{32}$ is proportional to the specific surface expressed as area per unit volume, the proportionality factor being 6 for spheres and cubes and greater than 6 for other shapes.

The weight mean diameter, $\bar{d}_{43}$, is called the De Brouckere diameter by Mugele and Evans.[21] Herdan[17] calls it the weight average particle size and defines it in general terms by the equation

$$\bar{d} = \Sigma d_i f_v(d_i^2)\, f_i/\Sigma f_v(d_i^2)\, f_i \tag{1-30}$$

The geometric mean is defined not by moments but by the equation for classified data:

$$\bar{d}_g = (d_1^{f_1} d_2^{f_2}\, d_3^{f_3} \cdots d_n^{f_n})^{1/n} \tag{1-31}$$

The populations of airborne particles are often log-normally distributed, as discussed later. The geometric mean of such a distribution is the value with the greatest frequency of particles, and is, thus, especially useful for calculations involving such a population.

Another mean that is sometimes used in aerosol research is the harmonic mean, which is defined by the equation

$$\bar{d}_h = \left[\frac{1}{n} \Sigma\, \frac{f_i}{d_i}\right]^{-1} \tag{1-32}$$

Table 1-3.   Some Definitions of Means.

Name	Symbol[a]	Definition
Arithmetic mean	$\bar{d}_{10}$	$\dfrac{1}{n}\,\Sigma d_i f_i$
Geometric mean	$\bar{d}_g$	$(d_1{}^{f_1}d_2{}^{f_2}d_3{}^{f_3}\cdots d_n{}^{f_n})^{1/n}$
Harmonic mean	$\bar{d}_{ha}$	$\left[\dfrac{1}{n}\,\Sigma\,\dfrac{f_i}{d_i}\right]^{-1}$
Mean surface diameter	$\bar{d}_{20}$ or $\bar{d}_s$	$\left(\dfrac{\Sigma f_i d_i{}^2}{n}\right)^{1/2}$
Mean weight diameter	$\bar{d}_{30}$ or $\bar{d}_w$	$\left(\dfrac{\Sigma f_i d_i{}^3}{n}\right)^{1/3}$
Linear mean diameter	$\bar{d}_{21}$ or $\bar{d}_1$	$\dfrac{\Sigma f_i d_i{}^2}{\Sigma f_i d_i}$
Surface mean diameter (also Sauter or volume surface)	$\bar{d}_{32}$ or $\bar{d}_{vs}$	$\dfrac{\Sigma f_i d_i{}^3}{\Sigma f_i d_i{}^2}$
Weight mean diameter (also DeBrouckere)	$\bar{d}_{43}$ or $\bar{d}_{wm}$	$\dfrac{\Sigma f_i d_i{}^4}{\Sigma f_i d_i{}^3}$

[a] When appropriate, the symbols are those corresponding to equation (1-26).

The arithmetic mean, geometric mean, and harmonic mean are related in the following way

$$\bar{d}_h \leq \bar{d}_g \leq \bar{d}_{10}$$

The inequalities increase with increasing spread of the size distributions. The diameters listed above are defined in Table 1-3.

2.6.   Distribution Functions and Plots

Perhaps the simplest type of graphical representation of a frequency distribution is the histogram. The abscissa represents the particle size and the ordinate the frequencies per class interval. The histogram is a bar graph in which the class interval is represented by the width of the bar and the frequency by the length of the bar. Figure 1-8 is the histogram constructed from the data of Table 1-1.

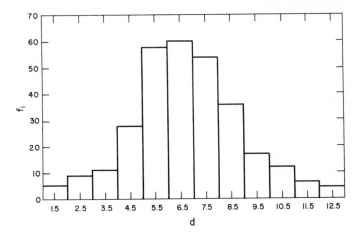

Figure 1-8.   Distribution of the diameters of 300 particles.

Now imagine that the midpoints of the tops of the bars of a histogram such as Figure 1-8 were connected by straight lines between adjacent points. By measuring an infinite number of particles and making the class intervals approach zero, the system of straight lines would become a smooth curve, with the ordinate representing a function, $f(d)$, of the particle size. The area under the curve and above any interval on the abscissa is the relative frequency with which the diameters of the particles occur in that interval. If the total area under the curve is normalized to unity, the relative frequency is the probability that a particle selected at random will have a size in the corresponding interval. The mathematical representation of the curve is known as the distribution function.

A distribution function that is commonly used for many types of statistical evaluations, but much less commonly for airborne particles, is the normal distribution. It is the frequency distribution produced when an infinite number of factors are introducing independent variations of equal magnitude, and is defined by the equation

$$f(d) = \frac{1}{s\sqrt{2\pi}} \exp\left[-\frac{1}{2}\left(\frac{d - \bar{d}}{s}\right)^2 \right]$$ (1-33)

where s, as usual, is the standard deviation. This distribution was first discovered by the English mathematician De Moivre (1667–1754) and was used extensively by the German scientist Gauss (1777–1855). It is often called the Gaussian distribution.

Very few particle size distributions, including those of airborne particles, are normal. Usually only aerosol particles produced by processes involving condensation have diameters that are normally distributed, and most airborne particles have skewed distributions. Often airborne particles have size distributions that can be satisfactorily represented by the log normal distribution. For such aerosols the logarithms of the diameters are normally distributed. For normally distributed diameters, if we select d_1 and d_2 such that they are on opposite sides of the arithmetic mean and equidistant from it, there is an equal probability that any randomly chosen particle would have the diameter d_1 or d_2. However, if the particles are log-normally distributed, the ratios of d_1 and d_2 to the mean, instead of the differences from the mean, must be equal in order to satisfy the condition of equal probability.

The log normal distribution is defined by the equation

$$ f(d) = \frac{1}{\log s_g \sqrt{2\pi}} \left[-\frac{1}{2} \left(\frac{\log d - \log \bar{d}_g}{\log s_g} \right)^2 \right] \qquad (1\text{-}34) $$

where $\bar{d}_g$ is the geometric mean and s_g is the geometric standard deviation. Note that $\log s_g$ is the standard deviation for the normally distributed logarithms. Some authors describe $\log s_g$ as the geometric mean standard deviation and represent it by the symbol s_g or σ_g.

Figure 1-9 shows three log-normally distributed distribution curves having the same value for the median, d_m, but differing values of s_g. Note that the curves are positively skewed (have a positive third moment).

When particles are log-normally distributed, the median is equal to the geometric mean. Other useful relationships are:

$$ \ln d_{m0} = \ln \bar{d}_g - \ln^2 s_g \qquad (1\text{-}35) $$

and

$$ \ln \bar{d}_{10} = \ln \bar{d}_g + 0.5 \ln^2 s_g \qquad (1\text{-}36) $$

where d_{m0} is the mode.

The size distributions of particles in the unpolluted atmosphere usually have a strong positive skew. Blifford and Gillette[22] investigated the application of the log normal distribution function for 329 such size distributions obtained from measurements of atmospheric aerosols collected at altitudes from the surface to 9 km. Using the Kolmogoroff-Smirnov test for accuracy of fit, they found that 98% were log normal when plotted as volume versus size distribution, assuming the geometric mean of each distribution to have values of 0.4–4.0 μm and s_g values from 1.26 to 30.0.

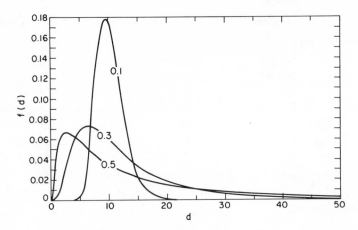

Figure 1-9. Three logarithmic normal distribution curves with $d_m = 10$ and $s_g = 0.1, 0.3,$ and 0.5. Based on A. Hald, *Statistical Theory with Engineering Applications*, Wiley, New York, 1962.

The size distributions of aerosol particles are sometimes represented by cumulative curves, which are produced by plotting the percentages of particles greater than (or less than) a given particle size against the particle size. Thus, the limiting values for the ordinate may vary from 0 to 100%. Of course, the ordinate can be chosen to represent the percentage of total surface or of total weight instead of the percentage of the number of particles. Like the size-frequency curves, cumulative curves have the advantage over histograms that the class interval is eliminated. Another advantage is that they can readily be prepared from classified data having unequal class intervals. They are especially useful for locating the median and quantile values.

When particle size distributions are normally or log-normally distributed, they can be represented by straight lines by plotting them in cumulative form on probability or log probability paper, respectively.

Probability paper is prepared in the following manner. Consider a cumulative distribution such that the number of particles less than d in diameter, R_n, is represented by an equation of the type

$$R_n = f_i(d) \tag{1-37}$$

Obviously, a straight line is obtained by plotting R_n versus $f_i(d)$. If the diameters are normally distributed, an expression in the form of equation (1-37) can be obtained by integrating the equation for the normal distribution (equation 1-33).

$$R_n = \frac{100}{s\sqrt{2\pi}} \int_0^d \exp\left[-\frac{1}{2}\left(\frac{d - \bar{d}}{s}\right)^2\right] d(d) \qquad (1\text{-}38)$$

The factor 100 is applied because cumulative curves are expressed in terms of percentages instead of probability. Equation (1-38) cannot be integrated directly but can be evaluated from tables of integrals. Probability paper represents d on a linear scale and R_n on a probability scale such that a straight line is obtained when equation (1-38) is satisfied. Log probability paper is correspondingly based on the integrated form of equation (1-34).

$$R_n = \frac{100}{\log s_g \sqrt{2\pi}} \int_0^d \exp\left[-\frac{1}{2}\left(\frac{\log d - \log \bar{d}_g}{\log s_g}\right)^2\right] d(\log d) \quad (1\text{-}39)$$

When any one of the size distributions number, surface, and weight is log-normally distributed, the other two are also log-normally distributed. For a given aerosol, all three distributions will have the same standard deviation and the straight-line plots will be parallel. Thus, when one distribution has been plotted, one point for each of the other distributions completely determines their cumulative distribution straight lines. Given the geometric mean diameter for one of these distributions and the geometric standard deviation, other mean diameters can be calculated from the following equations:

$$\ln \bar{d}_g = \ln \bar{d}_{wg} - 3 \ln^2 s_g \qquad (1\text{-}40)$$

and

$$\ln \bar{d}_{32} = \ln \bar{d}_{wg} - 0.5 \ln^2 s_g \qquad (1\text{-}41)$$

where $\bar{d}_{wg}$ is the geometric mean of the weight distribution. If one has reason to believe that a distribution is log normal, the geometric standard deviation can be calculated from the equalities

$$s_g = \frac{84.13\% \text{ size}}{50\% \text{ size}} = \frac{50\% \text{ size}}{15.87\% \text{ size}} \qquad (1\text{-}42)$$

The expression for the log-normal distribution must be integrated with respect to log d. Since the differential of ln d is $d(d)/d$, the expression for the log-normal distribution is sometimes written as

$$f(d) = \frac{1}{(2\pi)^{1/2} d \log s_g} \exp\left[\frac{(\log d - \log \bar{d}_g)^2}{2 \log^2 s_g}\right] \qquad (1\text{-}43)$$

which gives the population density correctly when integrated with respect to d.

Espenscheid et al.[23] have shown that equation (1-39) integrated with respect to d leads to a new distribution function which they call the zeroth-order logarithmic distribution function (ZOLD):

$$f(d) = \frac{\exp\left(-(\log d - \log d_{m0})^2/2 \log^2 s_0\right)}{\int_0^\infty \exp - (\log d - \log d_{m0})^2/2 \log^2 s_0 \, d(d)}$$

$$= \frac{\exp - (\log d - \log d_{m0})^2/2 \log^2 s_0}{\sqrt{2\pi} \, d_{m0} \log s_0 \exp (\log^2 s_0/2)} \tag{1-44}$$

This distribution is defined by the modal diameter, d_{m0} and by s_0, which is a measure of the width and skewness of the distribution. The modal value is related to the arithmetic mean value by the equation

$$\ln \bar{d} = \ln d_{m0} + 1.5 \ln^2 s_0 \tag{1-45}$$

The standard deviation is given by the equation

$$s = d_{m0}[\exp (4 \log^2 s_0) - \exp (3 \log^2 s_0)]^{1/2}$$

$$= d_{m0} \log s_0 \left(1 + \frac{7 \log^3 s_0}{2!} + \frac{37 \log^4 s_0}{3!} + \cdots \right)^{1/2} \tag{1-46}$$

thus for $\log s_0 < < 1$, $s = d_{m0} \log s_0$ $\qquad$ (1-47)

Earlier, the coefficient of variation was defined as the standard deviation expressed as a percentage of the arithmetic mean, $\bar{d}$. From equation (1-47) it is seen that $\log s_0$ has a similar significance, since

$$\log s_0 \cong \frac{s}{d_{m0}} \tag{1-48}$$

when $\log s_0 < < 1$.

Espenscheid et al.[23] point out that the logarithmic distributions described above are members of a general family of logarithmically skewed distributions having the form

$$f_n(d) = \frac{d^n \exp - (\log d - \log d_n)^2/\log^2 s_n}{\sqrt{2\pi} \, d_n^{n+1} \log s_n \exp[(n + 1)^2 \log^2 s_n/2]} \tag{1-49}$$

When $n = -1$, equation (1-49) reduces to the log-normal distribution, and when $n = 0$ it reduces to the ZOLD—hence the name zero order. Also, for every value of n there is a corresponding value for the diameter, d_n, which is an indicator of central tendency and can be maintained constant as $\log s_n$ is changed.

The above logarithmic distribution functions represent distributions that are always positively skewed; that is, the prepondence of the particles is larger than the mode. However, the distributions of airborne particles are not always either uniformly distributed about the mean or positively skewed.

The problem of variable skew has been investigated by Rowell and Levit.[24] The negatively skewed Wallace-Kratohvil distribution (WKD) is defined by

$$f_{wk}(d) = \frac{C_{wk} \exp\left(-\log^2[1 + (d_m - d)/d_{m0}]\right)}{2 \log^2 s_0}$$

where C_{wk} is the normalization constant and d_{m0} and s_0 are the same as for ZOLD. Rowell and Levit found C_{wk} to be the same as the normalization constant for ZOLD, namely

$$C_{wk} = 1/\sqrt{2\pi} \, d_{m0} \log s_0 \exp \frac{\log^2 s_0}{2} \qquad (1\text{-}50)$$

These authors also represented negatively skewed distributions by a modification of the Khrgian–Mazin function:

$$f(x) = C_{km} \, d^A \exp\left(-B \, d^A\right) \qquad (1\text{-}51)$$

where $B = 1/d_{m0}{}^A$. They normalized the distribution using the gamma function:

$$C_{km} = \frac{A/d_{m0}{}^{A+1}}{\Gamma(1 + 1/A)} \qquad (1\text{-}52)$$

The degree of skew was varied by varying A, but rather large values of A were required to produce negative skew in narrow size distributions.

As Rowell and Levit pointed out, it is certainly awkward to change from one function to another when a distribution changes from negative to positive skew, or the reverse. Therefore, they proposed two new functions, each of which can be used to represent either positively or negatively skewed distributions. The first of these they call the skewed normal distribution, SND:

$$f(x) = [(1 - A^2)/s\sqrt{2\pi}] \exp - [(d - d_{m0}) - A|d - d_{m0}|]^2/2s^2 \qquad (1\text{-}53)$$

where $-1 < A < 1$. The skew is negative or positive according to the sign of A, and reduces to the normal distribution when A is zero, in which case s, the breadth, becomes the normal standard deviation. As this implies, the skew is sensitive to the magnitude of A, since the skew is developed in the exponential term by taking from one side of the distribution and simultaneously adding to the other.

The skewed normal distribution, like the normal distribution, has the theoretical disadvantage of admitting negative values of the diameter. This, of course, cannot be the case for logarithmic distributions. Rowell and Levit proposed a logarithmic distribution that yields the ZOLD and WKD as special cases. They call this function the skewed zeroth-order logarithmic function, SZOLD:

$$f(x) = (A/s' \, d_{m0}\sqrt{2\pi} \, \exp \, (s')^2/2)x \, \exp \, - \, [\log^2(1 + A(d - d_{m0})/d_{m0}]/2(s')^2 \tag{1-54}$$

where s', to be consistent with the previous logarithmic functions, equals $\log s$. When A is $+1$ the function yields the positively skewed ZOLD, and when A is -1 it yields the negatively skewed WKD. The curves for these two special cases are mirror images of each other, and as A approaches zero the skew and breadth increase until the function vanishes when A equals zero. When s' is small and A is large, the SZOLD approximates a normal distribution.

Rowell and Levit also proposed a new approach for representing breadth and skew, which involves variables directly related to A and $\log s$ for SND and SZOLD. This approach is based on measuring the width, w, or displacement of the curve from the modal diameter at a specified height or fraction of the modal value of $f(x)$. For SZOLD the value of d_a when $f(x)$ has dropped to e^{-a} times the modal value of $f(x)$ is given by the equation

$$e^{-a} = \frac{\exp \, - \, \log^2[1 + A(d_a - d_{m0})/d_{m0}]}{2(s')^2\pi} \tag{1-55}$$

and

$$w = d_a - d_{m0} = \frac{d_{m0}}{A} \, [\exp \, (\pm s'\sqrt{2_a}) - 1] \tag{1-56}$$

The $\pm$ signs corresponds to displacements on either size of the mode.

They define a sigma width, w_σ, for a $= \frac{1}{2}$, that is, when the distribution has fallen to $e^{-1/2}$ (61%) of the modal value of $f(d)$, which for SZOLD is

$$w_\sigma = \frac{d_{m0}}{A} \, [\exp \, (\pm s') - 1] \tag{1-57}$$

To characterize the spread by the width at 50% of the mode (the half-band width), w_h:

$$w_h = \frac{d_{m0}}{A} \, (\exp \, (\pm s'\sqrt{2 \ln 2} - 1)$$

They also define the "base width" to characterize the width of the distribution as $A = 9/2$. For SZOLD this is given by the expression

$$w_b = \frac{d_{m0}}{A} \left(\exp \left(\pm 3s'\right) - 1\right)$$

The correponding widths for the SND are

$$w = \pm s\sqrt{2a}/(1 \pm A) \tag{1-58}$$

$$w_\sigma = \pm s/(1 \pm A) \tag{1-59}$$

$$w_h = \pm s\sqrt{2 \ln 2}/(1 \pm A) \tag{1-60}$$

$$w_b = \pm 3s/(1 \pm A) \tag{1-61}$$

The breadth measures for SND reduce to the normal distribution for $A = 0$.

The width parameters can be used to represent the extent of skew as the ratio of the width on the broad side of the mode to that on the narrow size. Rowell and Levit called this ratio the balance, b. For SZOLD the balance is independent of the skew parameter A but does depend on the breadth parameter, s':

$$b = [\exp (s'\sqrt{2a}) - 1]/[\exp (-s'\sqrt{2a}) - 1] \tag{1-62}$$

$$b_\sigma = [\exp (s') - 1]/[\exp (-s') - 1] \tag{1-63}$$

$$b_h = [\exp (s'\sqrt{2 \ln 2}) - 1]/[\exp (-s'\sqrt{2 \ln 2}) - 1] \tag{1-64}$$

$$b_b = [\exp (3s') - 1]/[\exp (-3s') - 1] \tag{1-65}$$

The balance for SND is independent of the frequency level, and depends only on the skew parameter, A:

$$b = (1 + A)/(1 - A) \tag{1-66}$$

A function that has often been used to represent the size distributions of powders, especially when the distributions are obtained by sieving and occasionally for suspensions of droplets, is the Rosin–Rammler relationship. It is often defined in the form

$$1 - v = \exp [-d/(d')^b] \tag{1-67}$$

where $1 - v$ is the volume fraction of particulate material occurring in particles having a diameter greater than d, d' is a size parameter, and

b is a distribution parameter. When equation (1-67) is differentiated, one obtains the volume-distribution form:

$$\frac{dv}{d(d)} = b\,\frac{d^{b-1}}{(d')^{b}}\,\exp\left(-\frac{d}{d'}\right)^{b}$$ (1-68)

Equation (1-68) can be written in number-distribution form by dividing by d^3 and inserting a factor to make

$$\int_0^{\infty}\frac{dN}{d(d)}\,d(d) = 1:$$

$$\frac{dN}{d(d)} = \frac{bd^{b-4}}{(d')^{b-3}\,\Gamma(1-3/b)}\,\exp\,(d/d')$$ (1-69)

The gamma function can be evaluated using tables of this function.

Mean diameters can be calculated from the equation

$$(\bar{d}_{qp})^{q-p} = (d')^{q-p}\,\Gamma\left(\frac{q-3}{b}+1\right)\Big/\Gamma\left(\frac{p-3}{b}+1\right)$$ (1-70)

where q and p are defined by equation (1-26). This equation should be used with caution, however; although it is theoretically correct, it can produce quite erroneous results for suspensions of particles that are fairly well represented by the Rosin-Rammler function.

This function is often used in the form

$$\ln\ln\frac{1}{1-v} = b(\ln d - \ln d')$$ (1-71)

which has the advantage that if a plot of the values of a size distribution corresponding to the two sides of equation (1-71) produce a straight line, the Rosin–Rammler function will provide at least a fairly good representation of the size distribution. Furthermore, the slope of the straight line produced by plotting (on log-log paper) log $1/(1-v)$ vs d is b, and d' is the value of d for which $1 - v = e^{-1}$.

Several entirely empirical distribution functions are used that have an exponential form, for example the Nukiyama–Tamasawa equation[25] which was developed for spray droplets:

$$\frac{dN}{d(d)} = Dd^2e^{-bd^c}$$ (1-72)

Equation (1-72) can be written in terms of a gamma function:

$$\frac{d(N)}{d(d)} = \frac{cb^{3/c}}{\Gamma(3/c)}\,d^2e^{-bd^c}$$ (1-73)

or for the volume distribution:

$$\frac{dv}{d(d)} = \frac{b^{6/c}}{\Gamma(6/c)} d^5 e^{-bd^c} \tag{1-74}$$

Mugele and Evans[21] showed that mean diameters can be calculated from the equation

$$\bar{d}_{qp}^{q-p} = b^{(q-p/c)} \Gamma\left(\frac{q-3}{c}\right) / \Gamma\left(\frac{p-3}{c}\right) \tag{1-75}$$

Two very simple functions that have been used extensively in atmospheric research, for both natural and polluted atmospheres, are

$$\frac{dN}{d(d)} = ad^{(c-1)} \tag{1-76}$$

and

$$\frac{dN}{d(\log d)} = \frac{d(dN)}{d(d)} = ad^c \tag{1-77}$$

The functions at best describe only part of the size distribution, since N becomes infinite at one end or the other of the distribution represented by the function, depending on the sign of c. Junge[26,27] found c to be about -3 over much of the size range of particles in the natural atmosphere. However, the value of c varies markedly as a function of time and altitude, as can be seen from Figures 1-10 and 1-11 where c is the slope of each curve as plotted on the log-log scale.[28]

When undertaking atmospheric research on particulate matter, one is usually interested in both the size distributions and the concentrations of particles in a given size range. Thus it is useful to plot dN/d (log d) or $dN/(d \log r)$ per cubic centimeter of air against d or r, the radius, as was done in Figure 1-10 and 1-11. The area under the curve then corresponds to the total number of particles per cubic centimeter rather than to the probability of finding a particle in the corresponding size range. If we now define

$$n(r) = \frac{dN}{d(\log r) \text{ cm}^{-3}} \tag{1-78}$$

the number of particles per cubic centimeter between the limits of the interval Δ (log r) may be approximated from the equation:

$$\Delta N = n(r) \, \Delta \log r \tag{1-79}$$

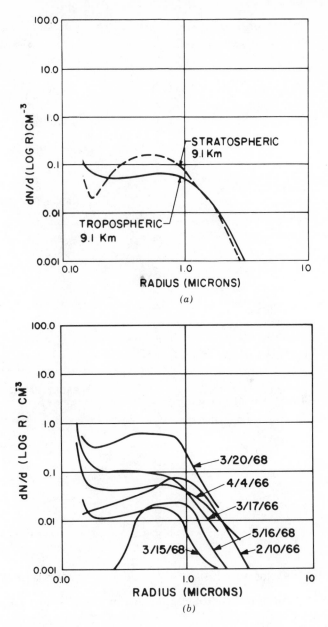

Figure 1-10. (a) Average $dN/d(\log r)$ [cm^{-3}] versus particle radius for collections made in stratospheric air and in the troposphere at 9.1 km; (b) $dN/d(\log r)$ [cm^{-3}] versus particle radius for individual collections made in stratospheric air at 9.1 km.[28]

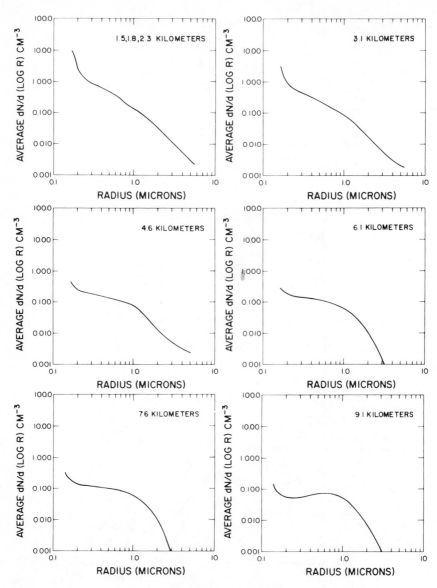

Figure 1-11.   Average particle size distributions at altitudes from 1.5 to 9.1 km over Nebraska.[28]

This equation is, of course, not exact, since $n(r)$ is changing throughout the interval, the exact equation being

$$\int_{N_1}^{N_2} dN = \int_{r_1}^{r_2} n(r) \, d \, (\log r) \tag{1-80}$$

The corresponding log radius–surface (S) and log radius–volume (v) distributions can be obtained from the equations

$$S(r) = \frac{dS}{d(\log r)} = \frac{4\pi r^2 \, dN}{d(\log r)} \tag{1-81}$$

and

$$v(r) = \frac{dV}{d(\log r)} = \frac{(4/3) \, \pi r^3 \, dN}{d(\log r)} \tag{1-82}$$

The log radius–mass distribution can be calculated from the equation

$$m(r) = \rho v(r) \tag{1-83}$$

where ρ is the particle density.

The size distributions of aerosol particles are often bimodal or even multimodal. Sometimes multimodal size distributions can be represented by the sum of two or more unimodal functions, but this approach is often unsatisfactory. Dalla Valle et al.[29] have proposed the following function for representing bimodal particle size distributions:

$$\frac{dN}{d(d)} = \exp \left[-a_4 d^4 + a_3 d^3 + a_2 d^2 + a_1 d + a_0)\right] \tag{1-84}$$

This can be put in the form

$$-\ln \frac{dN}{d(d)} = g(d) \tag{1-85}$$

and $g(d)$ in turn can be rewritten:

$$g(d) = a^2(d^4 + a_3' d^3 + a_2' d^2 + a_1' d + a_0') \tag{1-86}$$

where $a_3' = a_3/a^2$; $a_2' = a_2/a^2$, and so forth.

By replacing d by $d - a_3'/4$, equation (1-86) becomes

$$g(d) = a^2(d^4 + td^2 + ud + w) \tag{1-87}$$

where t, u, and w are constants. Then the bimodal distribution function is in applicable form:

$$\frac{dN}{d(d)} = \exp \left[-a^2 w\right] \exp \left[-a^2(d^4 + td^2 + ud)\right] \tag{1-88}$$

The values for the modes can be obtained by differentiating (1-88) and equating the result to zero:

$$4d^3 + 2td + u = 0 \tag{1-89}$$

When the roots of equation (1-89) are real and distinct they are the values for the two modes and the minimum between the modes. The roots are real and distinct when

$$-8t^3 > 27u^2 \tag{1-90}$$

Dalla Valle et al. suggested several methods for evaluating the constants from actual size distribution data.

3. THE PRODUCTION OF SIZE DISTRIBUTIONS

A knowledge of the manner in which various size distributions result from various methods of producing aerosols is useful in predicting size distributions before any measurements are made (knowing "what to expect"), as well as in interpreting the results of such measurements.

Many aerosols are prepared having as narrow a size distribution as is reasonably possible. These "monodispersed" aerosols are used to study aerosol properties and for the calibration of aerosol-measuring instruments. Conversely, a number of instruments for determining the average size of aerosol particles can only be used for nearly monodispersed aerosols. Numerous methods are available for preparing them,[30,31,32] including the condensation of droplets from the vapor phase, mechanical dispersion of liquids, and the dispersion of powders consisting of particles that are nearly all the same size. A discussion of such methods is beyond the scope of this book.

Junge,[33,34] as a result of many years of investigation of the size distributions of particles in the natural atmosphere, has suggested that on the average the distribution in continental air is like that shown in Figure 1-12. This has come to be known as the Junge distribution. Note that it differs somewhat in detail from the distributions found by Blifford and Ringer,[28] shown in Figures 1-10 and 1-11. In fact, the results of Blifford and Ringer for air in the lower stratosphere and upper troposphere suggest that there may be a bimodal distribution, and the results of Bigg[35] suggest a multimodal distribution for the stratosphere. Nonetheless, numerous attempts have been made to explain the apparent persistence of the Junge distribution and to represent it by mathematical "models."

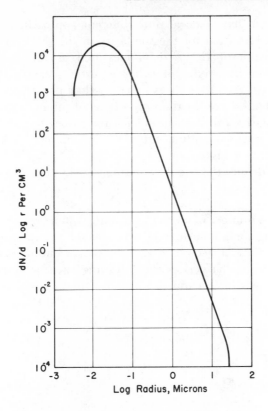

Figure 1-12.   Particle size distribution for continental air.[34]

Junge[33] has proposed a model based on the proposition that particles over the entire range of sizes covered by Figure 1-12 are distributed throughout the atmosphere by eddy diffusion. The size distribution is determined to a considerable extent by coagulation and sedimentation. Coagulation controls the lower size limit because below a certain size (really, size range) coagulation is so rapid that particles cease to exist as independent species almost immediately; coagulation also controls the size distribution of the "Aitken" particles (those less than 0.1 μm in radius). Sedimentation controls the upper size limit, and Junge suggests that this must be about a 20-μm radius. He also makes the rather interesting observation that the size of pollen grains should represent the upper limit of particles that can stay airborne for an appreciable length of time and that the pollen of a large variety of plants is very uniform in size, about a 10-μm radius.

Friedlander[36,37] has taken a quite different approach. He points out that Junge's explanation for the similarity in the shape of the size distributions of atmospheric particles measured at different times and places is based on the mixing of airborne particles having rather special size distributions. Friedlander's explanation does not depend on the initial size distribution nor on the proportions of mixing. His method involves a "similarity transformation" for the upper end of the particle size distribution based on the equations describing the kinetics of a coagulating, settling suspension.

Friedlander defines the size distribution $n(r)$ by the equation

$$dN = n \, dr \tag{1-91}$$

where dN is the number of particles per unit volume between r and $r + dr$. When using the Smoluchowski equation for the rate at which particles of class r' and r'' combine and the Stokes-Cunningham equation for settling velocity, he derives the following equation for the rate of change of the distribution function of a coagulating, settling aerosol:

$$\frac{dN}{dt} = \frac{K}{4} \int_0^4 \emptyset[r', (r^3 - r'^3)^{1/3}] \, n(r') \, n[r^3 - r'^3)^{1/3}]$$

$$\times \frac{r^2}{(r^3 - r'^3)^{2/3}} \, dr' - \frac{K}{2} \, n \int_0^\infty \emptyset(r,r') \, n(r') \, dr'$$

$$- vs \frac{dn}{dy} \tag{1-92}$$

where K = coagulation constant, $\emptyset (r', r'') = [(r + \alpha)/r'^2 + (r'' + \alpha)/r''^2] (r' + r'')$, α = constant $(\alpha/r$ = "Cunningham correction"), v_s = Stokes–Cunningham settling velocity = $Sr^2 (1 + \alpha/r)$, $S = 2g\rho/9\mu$, g = gravitational constant, ρ = particle density, μ = gas viscosity and y = distance along the gravitational field.

Although, in principle, this equation could be solved, given appropriate initial and boundary conditions, it is unlikely that such information will ever be available. Therefore, Friedlander used another approach. The persistence of the size distribution in the atmosphere suggests that a local dynamic equilibrium is maintained between the formation of large particles by coagulation and their removal by sedimentation. Possibly, the depletion from the Aitken range is made up by the diffusion of Aitken particles from the surrounding air, or if the depletion is not made up, there is a slow change in concentration with time. The decay of turbulence in a fluid behaves in a very similar way; thus Friedlander applied a "similarity transformation" to equation (1-

92) comparable to the one that had been applied to the corresponding equation for the decay of turbulence. The result was the following equation, which neglects the Cunningham correction for the upper end of the size distribution:

$$\frac{3\eta^5}{32\pi I}\frac{d(\psi/\eta^2)}{d\eta} + \frac{\eta^2}{4}\int_0^\eta \left(\frac{1}{\eta'} + \frac{1}{\eta''}\right)\frac{(\eta' + \eta'')}{\eta''^2}$$

$$\times \psi(\eta')\,\psi(\eta'')\,d\eta' - \frac{1}{2}\int_0^\infty \left(\frac{1}{\eta'''} + \frac{1}{\eta}\right)(\eta''' + \eta)$$

$$\times \psi(\eta''')\,\psi(\eta)\,d\eta''' = 0,$$

and

$$(\eta'^3 + \eta''^3) = \eta^3 \tag{1-93}$$

where ψ is a dimensionless function, $\eta = r/l_1$, l_1 is a characteristic spectral length, and $= \epsilon^{1/9} K^{2/9} S^{-1/3}$, ϵ is the rate at which matter enters the upper end of the size distribution by coagulation, and

$$I = \int_0^\infty \psi\,\eta^5\,d\eta$$

The primes on η refer to different classes of particles.

Although it may be possible to solve equation (1-93) with the boundary conditions $\psi = \psi_0$ at $\eta = \eta_0$ (the lower end of the distribution where equilibrium still occurs) and $\psi = 0$ at $\eta = \infty$, Friedlander did not attempt to do so. Instead, by making an assumption concerning the nature of the transfer process, he was able to derive specific forms for the dependence of n on r. In the size range $0.5 > r > 0.1$ μm, sedimentation was assumed to be negligible, and the only change that occurs is coagulation. Since the concentration in this range is orders of magnitude greater than that above 1 μm, nearly all the matter being transferred up the size distribution will pass through this range at the rate ϵ, and the distribution function will depend on r, ϵ, and K but not on S. Thus by dimensional analysis,

$$n = A_1 \left(\frac{\epsilon}{K}\right)^{1/2} r^{-5/2} \tag{1-94}$$

where A is a constant. At the upper end of the size range there may be a region in which coagulation is unimportant, n now being a function only of r, ϵ, and S:

$$n = A_2 \left(\frac{\epsilon}{S}\right)^{3/4} r^{-19/4} \tag{1-95}$$

Friedlander emphasizes that his explanation requires no assumption concerning the nature of the size distribution of the particles introduced into the atmosphere, nor any conjectures about the atmospheric boundary conditions. It is the so-called self-preserving hypothesis, because the reduced distribution function does not depend explicitly on time.

Friedlander[38] used a similar approach for the size distribution of the stratospheric aerosol. The main difference from the development he used for the tropospheric case was that a similarity transformation was applied to the lower end of the size distribution, which was defined as that part of the distribution for which sedimentation can be neglected. It was assumed that the distribution function $n(r)$ vanishes as r approaches zero, reaches a peak value, and then decreases rapidly in the equilibrium region. An equation for the decay with time of the total particle concentration N was derived:

$$\frac{1}{N^2} \sim t \tag{1-96}$$

Brock[39,40] investigated theoretically some of the factors influencing the size distribution of atmospheric aerosols. He assumed that particles in the size range > 0.1 μm are being continually supplied both by growth by condensation from the small end of the size distribution and by "primary sources" throughout the size distribution. He also assumed that dispersion, mixing, sedimentation, scavenging, and impaction all removed particles from this size range. He demonstrated that these processes would produce the characteristic skewed, long-tailed distributions, regardless of the size distributions of the particles in the primary sources.

Blifford[41] has worked with Junge, making modifications in the latter's model and taking advantage of the modern very large computers. The model is shown schematically in Table 1-4. The model was thought of as a global steady-state model, and the fraction of the aerosol in the clouds as well as that outside was considered. Thermal coagulation was treated for the fraction outside the clouds and included sedimentation and shear flow coagulation in the coagulation term. However, Blifford used finite difference equations instead of integrodifferential equations as used by Friedlander. Blifford's treatment also took into account various chemical reactions that may be proportional to the surface area of the particles, such as the surface-catalyzed photooxidation of sulfur dioxide. Also included was washout by rain, sedimentation, and attachment to leaves and other surfaces.

The second part of the model relates to the portion of the aerosol within clouds. This also included two parts. The first was the particulate

Table 1-4.   Processes Considered in the Computation of Tropospheric Aerosol.[41]

$$n(r) = n_1(r) + n_2(r) + n_3(r)$$

	Fraction inside clouds = η $n_2(r) + n_3(r) = \eta \cdot n(r)$	
Fraction outside clouds = $(1 - \eta)$ $n_1(r) = (1 - \eta)\cdot n(r)$	Fraction activated = κ $n_2(r) = \kappa\cdot\eta\cdot n(r)$	Fraction not activated = $(1 - \kappa)$ $n_3(r) = (1 - \kappa)\cdot\eta\cdot n(r)$
Thermal coagulation	Coagulation of cloud droplets	Thermal coagulation[a]
Sedimentation coagulation	Thermal coagulation of particles with cloud droplets[a]	Thermal coagulation of particles with cloud droplets[a]
Shear flow coagulation	Sedimentation coagulation of particles with cloud droplets[a]	Heterogeneous gas reactions
Homogeneous gas reactions	Diffusiophoresis	
Heterogeneous gas reactions	Heterogeneous gas reactions	
Washout	Rainout	
Sedimentation		
Attachment to obstacles		

[a] Corrected for particle growth at high humidity.

material that Blifford called the activated fraction and consisted of particles that serve as condensation nuclei. The processes for this portion include the coagulation of cloud droplets, the agglomeration of cloud droplets with particles, the impaction of particles by sedimenting cloud droplets, diffusiophoresis, heterogeneous gas reactions, and rainout. Rainout can be defined as all processes within the clouds that ultimately result in the removal of particles by rain. The second part of the aerosol within the clouds consists of particles that are not activated, and the processes again include agglomeration with each other and with cloud droplets and heterogeneous chemical reactions involving gases.

Hidy has also reviewed the various models and proposed modifications of his own.[42,43]

None of the models gives a perfect fit to the "Junge distribution," and in any case the Junge distribution is merely the average of many size distributions. This distribution may represent a portion of a log normal distribution. Thus the agreement of the various models with the average size distributions for atmospheric particles may in a sense be fortuitous, the models representing in a very imperfect manner the processes actually occurring in the atmosphere. Nonetheless, the attempts to formulate such models have induced various scientists to examine in detail the processes affecting the size distributions of atmospheric particles.

There are numerous mechanisms that can produce polymodal size distributions in aerosols, and those mechanisms usually involve two or more different sources for the particles. Preferential removal of a given size region can also produce a polymodal size distribution. Bimodal distributions can be observed in photochemical smog which may result from the addition of the particulate products of the photochemical reactions to the other components of the smog. Another mechanism has been proposed by Sutugin et al.[44] Monodispersed aerosols with low particle concentration can at times be formed by the condensation of some inorganic substances. The condensation is entirely heterogeneous, occurring on some foreign nuclei. However, Sutugin et al. demonstrated that under some conditions, especially high concentrations of the vapor, a bimodal distribution is obtained, apparently as a result of simultaneous homogeneous and heterogeneous nucleation. The study was both theoretical and experimental.

4. REPORTING CONCENTRATIONS

Presenting the results of concentration measurements is usually much simpler than presenting size distribution data. One method for presenting such information has already been discussed, namely plotting dN/dr per cubic centimeter or $dN/d \log r$ per cubic centimeter against r. Of course, instead of dN, one could use dv or dm, making the plot on a volume or mass basis.

The usual presentation method is simply to state the number, volume, or mass of particles per unit volume of aerosol (gas plus particles). The units milligrams (mg), micrograms (μg), and even nanograms (ng) per cubic meter or per cubic centimeter are often used. However, even such a simple method for reporting results can lead to ambiguities. The most common ambiguity, as mentioned earlier, occurs when the results of concentration measurements made at high altitudes in the atmosphere are reported. At times such results are reported in terms of unit volumes at 760 torr and 273°K instead of ambient conditions. Authors tend to assume that the reader knows this has been done; thus if there is any possibility of ambiguity, some indication of the pressure and temperature of the aerosol in terms of which the concentrations are reported should be stated.

Another common method of reporting mass concentrations is the use of the concept of mixing ratio, namely mass of particles per unit mass of aerosol. This has the distinct advantage in atmospheric research of avoiding the difficulty arising from the dependence of temperature and pressure on altitude. The units used are often parts per million by weight (ppmw) or parts per billion by weight (ppbw).

REFERENCES

1. Cadle, R. D., *Particle Size Determination,* Interscience, New York, 1955.
2. Irani, R. R., and C. F. Callis, *Particle size: Measurement, Interpretation, and Application,* John Wiley & Sons, New York, 1963.
3. Orr, C., Jr., and J. M. DallaValle, *Fine Particle Measurement. Size, Surface, and Pore Volume,* Macmillan, New York, 1959.
4. Smales, A. A., Ed. *Particle Size Analysis,* The Society for Analytical Chemistry, London, 1967.
5. Cadle, R. D., *Trans. Am. Geophys. U., EOS,* **53,** 812 (1972).
6. Lodge, J. P., Jr., and E. R. Frank, *J. Micros.,* **6,** 449 (1967).
7. Cadle, R. D., W. H. Fischer, E. R. Frank, and J. P. Lodge, Jr., *J. Atmos. Sci.,* **25,** 100 (1968).
8. Cadle, R. D., and E. R. Frank, *J. Geophys. Res.,* **73,** 4780 (1967).
9. Martin, G., C. E. Blythe, and H. Tongue, *Trans. Ceram. Soc. (Eng.)* **23,** 61 (1924).
10. Tomkeiff, S. L., *Nature,* **155,** 24 (1945).
11. Moran, P. A. P., *Nature,* **154,** 490 (1944).
12. Feret, L. R., *Assoc. Int. pour. l'Essai des Mat.* **2,** Group D, Zurich (1931).
13. Cadle, R. D., and W. C. Thuman, *Ind. Eng. Chem.,* **52,** 315 (1960).
14. Heywood, H., The scope of particle size analysis and standardization, in Symposium on Particle Size Analysis, Institution of Chemical Engineers, London, February 4, 1947.
15. Hausner, H. H., Characterization of the powder particle shape, in Smales, A. A., Ed., *Particle Size Analysis,* The Society for Analytical Chemistry, London, 1967.
16. Wadell, H., *J. Geol.,* **4,** 310 (1932).
17. Herdan, G., *Small Particle Statistics,* 2nd ed., Academic Press, New York, 1960.
18. Fisher, R. A. *Statistical Methods for Research Workers,* 13th ed., Hafner, New York, 1958.

19. Hoel, P. G., *Introduction to Mathematical Statistics,* Wiley, New York, 1947.

20. Wallis, W. A., and H. V. Roberts, *Statistics. A New Approach,* The Free Press, New York, 1956.

21. Mugele, R. A., and H. D. Evans, *Ind. Eng. Chem.,* **43,** 1317 (1951).

22. Blifford, Jr., I. H., and D. A. Gillette, Water, *Air and Soil Pollution,* 1, 106 (1971).

23. Espencheid, W. F., M. Kerker, and E. Matijivic, *J. Phys. Chem.,* **68,** 3093 (1964).

24. Rowell, R. L., and A. B. Levit, *J. Coll. Interface Sci.,* **34,** 585 (1970).

25. Nukiyama, S., and Y. Tanasawa, *Trans. Soc. Mech. Engrs. (Jap.),* 4, No. 14, 86 (1938).

26. Junge, C. E., *Tellus,* **5,** 1 (1953).

27. Junge, C. E., *J. Meteor.,* **12,** 13 (1955).

28. Blifford, Jr., I. H. and L. D. Ringer, *J. Atmos. Sci.,* **26,** 716 (1969).

29. Dalla Valle, J. M., C. Orr, Jr. and H. G. Blocker, *Ind. Eng. Chem.,* **43,** 1377 (1951).

30. Green, H. L., and W. R. Lane, *Particulate Clouds. Dusts, Smokes and Mists,* 2nd ed., Van Nostrand, New York, 1964.

31. Fuchs, N. A., and A. G. Sutugin, in G. M. Hidy and J. R. Brock, Eds., *Topics in Current Aerosol Research,* vol. 2, Permagon, New York, 1971.

32. Fuchs, N. A., and A. G. Sutugin, in C. N. Davies, Ed., *Aerosol Science,* Academic Press, New York, 1966.

33. Junge, C. E., *Air Chemistry and Radioactivity,* Academic Press, New York, 1963.

34. Junge, C. E., in U. S. Air Force, Handbook of Geophysics, Macmillan, New York, 1960.

35. Bigg, E. K., Personal communication, 1973.

36. Friedlander, S. K., *J. Meteor.,* **17,** 373 (1960).

37. Friedlander, S. K., *J. Meteor.,* **17,** 479 (1960).

38. Friedlander, S. K., *J. Meteor.,* **18,** 753 (1961).

39. Brock, J. R., *Atmos. Environ.,* **5,** 833 (1971).

40. Brock, J. R., in G. M. Hidy, Ed., *Aerosols and Atmospheric Chemistry,* Academic Press, New York, 1972.

41. Blifford, I. H., Jr., in I. H. Blifford, Jr., Ed., *Particulate Models: Their Validity and Application.* National Center for Atmospheric Research, Technical Note Proc. 68, Boulder, Colorado, 1971.

42. Hidy, G. M. ibid.

43. Hidy, G. M. and J. R. Brock, *The Dynamics of Aero-colloidal Systems,* Pergamon, New York, 1970.

44. Sutugin, A. G., A. A. Lushnikov, and G. A. Chernyaeva, *Aerosol Sci.,* **4,** 295 (1973).

2

COLLECTION WITHOUT
CLASSIFICATION BY SIZE

1. ADVANTAGES AND DISADVANTAGES

Airborne particles can be characterized either by first collecting them on some surface or by making measurements on them without removing them from the gas phase. Furthermore, if the particles are collected, the collection technique may be such that the particles are separated according to size, and the size distribution can be determined by measuring the amount of particulate material in each of the differently sized fractions collected. There are a number of rather obvious advantages and disadvantages to each of these approaches. The biggest advantage to collecting the particles is that if they are readily available for direct observation, a much more complete characterization is possible than if they are studied in the gas phase. Examples are collection by impaction, electrostatic precipitation, or thermal precipitation. A tremendous array of techniques is available for examining the individual particles, the bulk material (such as microanalytical techniques), or both. Even if the collected particles are not directly accessible (e.g., when they are collected on fiber filters), their concentration in the aerosol from which they were collected can be determined by weighing the filter before and after the collection or by quantitatively analyzing the collected material. Techniques are also available for removing the particles from filters in which they are imbedded, although for many types of particles (droplets, e.g.,) these techniques are unsatisfactory.

Separation into size fractions by the collection technique (e.g., by the cascade impactor) provides considerable direct information concerning the size distribution. However, since the size ranges overlap for most such techniques, interpretation of the results can be rather involved. Other disadvantages of collection prior to some sort of analysis are the following:

1. The particles may be changed by the collection technique. If they are droplets they may spread on the surface or even react chemically with the surface. If they are solid they may shatter, especially if they are collected by impaction.

2. Prior collection techniques are not readily adaptable to continuous, automatic measurement. They have been adapted to such measurements for very specific purposes (e.g., for measuring the "coefficient of haze" which is measured photometrically on a moving tape on which particles are deposited), but the results are far from satisfactory. Of course, after the particles have been collected the sizing and counting can be automated.

3. An additional source of error is introduced by undertaking the measurements in two steps instead of one.

Some of the advantages and disadvantages of direct measurement, without collection of the particles, are implied in the above discussion. Others are discussed later. In this chapter collection without classification by size is considered.

2. FILTRATION

Filtration is one of the oldest methods for collecting aerosol particles for some sort of characterization, and for many purposes it is still one of the best. Furthermore, the filtration of gas-borne particles has many industrial applications. These industrial filters range in sophistication and collection efficiency from the bags in "bag houses" to the "absolute" filters used in "clean rooms." Thus there is little wonder that a great deal of attention has been given to the theory of filtration and to the development of methods for designing filters for specific purposes.

Filters for aerosol characterization can be classified into two main types, namely, fiber filters and membrane filters. Fiber filters (sometimes called depth filters) can be prepared that are very efficient for a very wide particle size range and that have a very low pressure drop across the filter. However, they have the disadvantage that the collected particles are not readily accessible, for microscopy, for instance. The membrane

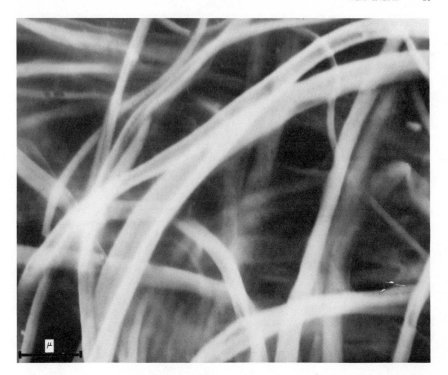

Figure 2-1.   Electron micrograph of a polystyrene fiber filter. Magnification 15,960×.

filters are small plastic sheets penetrated by extremely small and regularly sized holes. They have the advantage that most of the collected particles, and all of those larger than the holes, are collected on the surface of the sheet. A disadvantage is the rather high pressure drops. Granulated filters are also occasionally used. Figure 2-1 is an electron micrograph of a polystyrene fiber filter, Figures 2-2 and 2-3 are electron micrographs of two kinds of membrane filters, namely, "millipore" filters and "nuclepore" filters.

2.1.   Filtration Theory

2.1.1.   Fiber Filters

The theory of filtration, especially filtration by fiber filters, has been reviewed by Chen,[1] Dorman,[2,3] Pich,[4] and Davies.[5] The usual approach

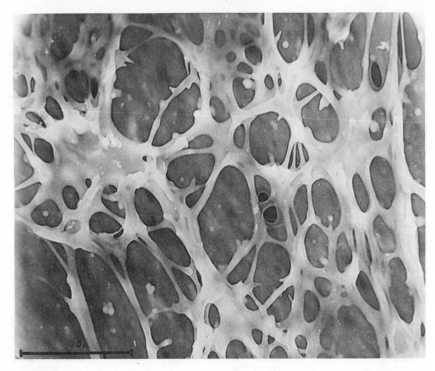

Figure 2.2. Electron micrograph of a "millipore" filter that had been impregnated with lead sulfide. Magnification 59,375×.

to treating filtration by fiber filters is first to consider the deposition by each of several mechanisms on single fibers, second to consider the combined mechanisms for single fibers, and third to apply the results for single fibers to entire filters. The properties of filters that are usually considered in theoretical treatments are usually the collection efficiency (E) and the pressure drop across the filter (ΔP). Of course, these are not the only properties of filters that must be considered when selecting filters for aerosol particle collection. Other properties include the strength of the filter, its chemical resistivity, and its ability to maintain physical integrity at high temperatures. The independent variables for which effects on E and ΔP are usually investigated are the size of the particles, the size of the fibers, and the flow rate of the gas. Other important considerations are the gas temperature, pressure, viscosity, and humidity; particle size distribution, shape, density, concentration, chem-

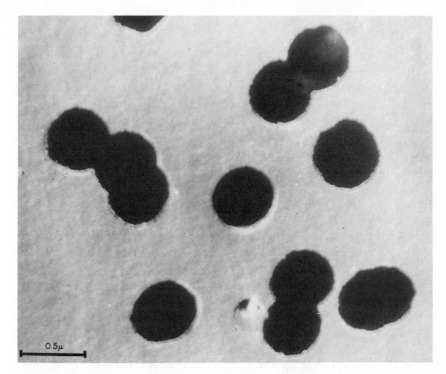

Figure 2-3.   Electron micrograph of a nuclepore filter surface. Magnification 33,120×.

ical constitution, and electrical charge; and fiber shape, chemical constitution, electrical charge, and size distribution.

Clearly, in the initial theoretical treatment all these variables cannot be treated simultaneously, and various simplifying assumptions must be made. These include the assumptions that the particles are spherical, that the fibers are circular in cross section, that each collision of a particle with a fiber results in adherence, and that the fibers are perpendicular to the direction of flow. Additional simplifying assumptions often made when filters rather than single fibers are considered are that the filters are not compressed by the pressure of the flowing aerosol and that the particle collection is not affected by the particles already collected by the fibers.

The main mechanisms of filtration under the conditions of viscous flow are direct interception, impaction of the particles on the fibers (inertial effects), and diffusion as a result of Brownian motion. Gravitational settling and electrical charges on the fibers, the particles, or both can also be involved.

Direct interception occurs when a particle following the flow lines of the air around the fiber comes in contact with it. If the radius of the fiber is a and of the particle is r, contact obviously occurs when the particle center reaches within $a + r$ of the fiber center, thus the likelihood of interception of the particle as the airstream is diverted around the fiber increases with increasing particle diameter.

Because of its inertia, an aerosol particle will not precisely follow the streamlines of gas flow as they are diverted around a fiber but will tend to continue to some extent in the direction it was going before the gas was diverted. A result is that many particles are impacted on the fibers that would not otherwise be collected. The relative effectiveness of this mechanism increases with increasing particle size and decreasing fiber diameter.

The Brownian motion (diffusion) of particles as they flow past a fiber also diverts them from the streamlines and can result in the particles reaching the fibers. This mechanism is especially effective for very small particles, and the use of fiber filters is one of the most effective means for collecting submicron-size particles from aerosols.

The development of equations representing the behavior of fiber filters begins by developing equations representing the velocity field around a single fiber, using the Navier–Stokes equations. These equations are only applicable when the gas phase of the aerosol can be considered to be continuous, that is, when the value of the Knudsen number (λ/D) is very small (λ = mean free path of the gas molecules and D = fiber diameter or thickness). Pich[4] states that for aerosols at atmospheric pressure this requirement is fulfilled for fibers about 10 μm in diameter or larger. However, the fiber filters used for sampling airborne particles are usually much thinner than this. Therefore, he considers four different flow regions. One, which he calls the classical region, is for $0 \leq \mathrm{Kn} \leq 10^{-3}$, where the equations of classical hydrodynamics apply. The second region, for $10^{-3} \leq \mathrm{Kn} \leq 0.25$, covers the flow of low-pressure gases. Pich calls the region $0.25 < \mathrm{Kn} < 10$ a transition region (the third region), and the fourth is in the region where $\mathrm{Kn} > 10$ and the gas flow is of a "molecular character." The velocity field for the second region can be calculated using either the equations of motion for low-pressure gases or the equations for classical hydrodynamics but with different boundary conditions, corresponding to the slip of gas at the fiber surface. Until the last few years only the theory of the classical region was studied to an appreciable extent, and most of the following discussion relates to that region.

Some of the earliest work in this field was undertaken by Lamb[6] who derived the following equations in polar coordinates for the flow (in two dimensions) for a nonviscous, irrotational, incompressible fluid perpendicular to an infinite cylinder of radius R:

$$v_r = v_0 \left(\frac{1 - R^2}{\psi^2} \right) \cos \theta \tag{2-1}$$

$$v_\theta = -v_0 \left(\frac{1 + R^2}{\psi^2} \right) \sin \theta \tag{2-2}$$

where ψ and θ are polar coordinates, v_r and v_θ are the velocity components in these coordinates, and v_0 is the upstream fluid velocity in the direction $\theta = 0$. Such flow is called potential flow. This velocity distribution holds for rather high values of the Reynolds number:

$$\text{Re} = \frac{D v_0 \rho}{\eta} \tag{2-3}$$

where ρ is the gas density and η is the viscosity of the gas.

Chen[1] points out that at low Reynolds numbers the velocity field around a fiber depends mainly on viscous forces that are absent in the case of ideal potential flow, assumed above, and that equations (2-1) and (2-2) would not be expected to apply to most filtration conditions. Lamb obtained the following approximate equations which can be expected to apply near the cylinder surface when the Reynolds number is less than unity:

$$v_r = C \left(1 - \frac{R^2}{\psi^2} - 2 \ln \frac{\psi}{R} \right) \cos \theta \tag{2-4}$$

$$v_\theta = C \left(1 - \frac{R^2}{\psi^2} + 2 \ln \frac{\psi}{R} \right) \sin \theta \tag{2-5}$$

where

$$C = \frac{v_0}{2(2 - 2 \ln \text{Re})} \tag{2-6}$$

Lamb obtained the following equation for the total drag force per unit length of a cylinder:

$$F = 8 \pi \eta C \tag{2-7}$$

The collection efficiency of a single cylindrical fiber, E, can be defined as the ratio of the cross-sectional area of the stream at large distances from the fiber from which particles are removed to the projected area of the fiber in the direction of flow.

Collection by impaction is generally expressed in terms of the inertial parameter, K, which is defined as follows for cylindrical fibers:

$$K = \frac{v_0 \rho_p \mathrm{d}^2}{18\eta D} \qquad (2\text{-}8)$$

where ρ_p is the particle density. The K is the ratio of the distance that a particle with initial velocity v_0 in still air will travel before coming to rest (the stopping distance) to the fiber diameter. Numerous investigators have studied the relationship between E_I (the collection efficiency for impaction) and K, usually presenting the results graphically. Landahl and Herrmann[7] are stated by Pich to have obtained the following empirical equation for Re = 10:

$$E_I = \frac{K^3}{K^3 + 0.77K^2 + 0.22} \qquad (2\text{-}9)$$

The results of several such calculations are shown in Figure 2-4. Note that some investigators have found a critical value of K below which no collection occurs.

These curves are for high Reynolds numbers. The curves for lower values of Re fall below those in Figure 2-4. Davies and Peetz[12] calculated E_I for several values of Re. Ranz and Wong[8] undertook an experimental determination of the efficiency curves for impaction by inertia

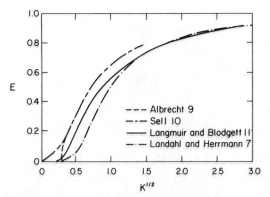

Figure 2-4. Theoretical impaction efficiencies of cylindrical collectors. Reprinted with permission from Ranz, W. E., and J. B. Wong, *Ind. Eng. Chem.*, **44**, 1371 (1952). Copyright by the American Chemical Society.

on cylindrical collectors. The aerosol material was concentrated sulfuric acid. The aerosol generator was of the type developed by Sinclair and his co-workers at Columbia University.[13] It involved condensation of the vapor of the aerosol material on nuclei under carefully controlled conditions, and it produced nearly monodispersed aerosols. The particle size and rate of generation were fixed by controlling the operating conditions. The particle size range was 0.36–1.3 μm diameter. A 77-μm diameter platinum wire was used as the cylindrical collector. The amount of aerosol collected was determined by washing the wire with distilled water and determining its electrical conductivity.

The results are shown in Figure 2-5. Ranz and Wong pointed out that the curve has an S shape, indicating a tendency to separate the size distribution for an aerosol into two size ranges. The results also suggest a minimum value of K below which particles are not collected. The general shape and relative position of the theoretical curves are confirmed; the theoretical curve of Langmuir and Blodgett[11] for low efficiencies is close to the experimental curve but rises above the experimental curve for intermediate values of efficiency.

Because the ratio of particle diameter to collector diameter was never greater than 0.017, impaction by interception was considered to be negligible.

Other experimental determinations of collection efficiency by single fibers have been made by Landahl and Herrmann,[7] Gregory,[14] and Wong et al.[15] Dorman[3] points out that all these experiments, which were

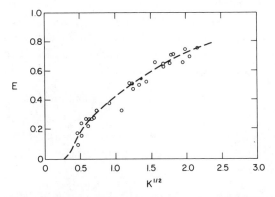

Figure 2-5. Experimental impaction efficiencies of cylindrical collectors. Reprinted with permission from Ranz, W. E., and J. B. Wong, *Ind. Eng. Chem.* **44,** 1371 (1952). Copyright by the American Chemical Society.

undertaken for medium and high values of Re, generally agree with the theory, but that there do not seem to be any reliable data for viscous flow.

Direct interception can be described in terms of the dimensionless parameter $J = d/D$ where d, as usual, is the diameter of the particle and D is the fiber diameter. The relationship of J to the collection efficiency obviously depends largely on the extent to which the particles follow the flow lines of the gas around the fiber. When the inertial parameter K approaches infinity, the particles are not diverted at all around the fiber, because of their great inertia, and the efficiency, E_R, equals J. On the other hand, when $K = 0$ and the particles have no inertia, they precisely follow the flow lines of the gas. Then, for potential flow Fuchs[16] gives the following equation:

$$E_R = 1 + J - \frac{1}{1 + J} \qquad (2\text{-}10)$$

For viscous flow the following equation was developed by Langmuir:[17]

$$E_R = \frac{1}{2(2 - \ln \mathrm{Re})} \left[2(1 + J) \ln (1 + J) - \frac{J(2 + J)}{1 + J} \right] \qquad (2\text{-}11)$$

Langmuir developed the following equation for the efficiency of a single fiber to collect particles by diffusion:

$$E_D = \frac{1}{2(2 - \ln \mathrm{Re})} \left[2(1 + x) \ln (1 + x) - (1 + x) + \frac{1}{1 + x} \right] \qquad (2\text{-}12)$$

where $x = 1.308 \, [(2\text{-}\ln \mathrm{Re})^{1/2}/\mathrm{Pe}^{1/2}]$, in which Pe is the Peclet number, Dv_0/D', and D' is the diffusion coefficient of the particles. Note the similarity between equations (2-11) and (2-12). The Peclet number characterizes diffusion deposition, much as J and K characterize interception and impaction, respectively. Ranz[18] considered diffusion to be analogous to heat transfer and obtained the equation:

$$E_D = (\pi \, \mathrm{Pe}^{-1}) \left[\frac{1}{\pi} + 0.55 \left(\frac{\eta}{\rho D'} \right)^{1/3} \mathrm{Re}^{1/2} \right] \qquad (2\text{-}13)$$

Chen[1] refers to Stairmond as using the same approach taken by Langmuir to calculate the diffusion collection efficiency for the ideal flow region. He developed the equation

$$E_D = 2\sqrt{2} \, \mathrm{Pe}^{-1/2} \qquad (2\text{-}14)$$

Langmuir[17] also investigated the problem of the efficiency of collection of particles by single fibers by the combined mechanisms of diffusion and interception, and Davies[19] has developed a theory for the combined collection due to inertia and interception.

Application of the theory described above to actual filters has not been very successful. One reason is that the fibers always vary in size, and the "effective mean diameter" will probably differ for the different collection mechanisms. Another reason is that the paths of flow among the fibers is tortuous, so that the flow field is difficult to define. Langmuir[17] developed the following equations on the assumption that only diffusion and interception effect the collection, that is, for low flow rates:

$$\frac{\Delta P}{\alpha \beta t} = \frac{2.24 \eta D'}{(\epsilon_0 + \epsilon_\lambda)(\epsilon_0 - \epsilon_S)^2 R^3} \tag{2-15}$$

and

$$\gamma = \frac{\epsilon_0 R A R'(\epsilon_0 + 2\epsilon_\lambda)}{\pi \eta} \tag{2-16}$$

where ΔP is the pressure drop, α is a constant near unity to correct for inaccuracies in the theory, β is the ratio fiber volume/filter volume, t is the filter thickness, r is the particle radius, $\epsilon_S = r/R$, $\epsilon_\lambda = 0.68\lambda/R$, A is the filter area, R' is the resistance per unit area and is equal to ΔP/volume flow rate of air, and $\epsilon_0 R$ is the minimum distance from the fiber of the layer of the aerosol from which particles diffuse to within a distance r of the cylinder. The filtering action, γ, is defined by the equation

$$\frac{N_1}{N_0} = \exp{(-\gamma)} \tag{2-17}$$

where N_0 is the number concentration of particles approaching the filter and N_1 is the number concentration in the air leaving the filter.

The value of ϵ_0 can be calculated from experimentally determined values of γ and AR' using the equation

$$\gamma = \epsilon_0{}^2 R A R'/\pi \eta \tag{2-18}$$

Langmuir also suggested that R could be calculated using the equation

$$R^2 = \frac{4M\phi t \beta L(1 - \beta) v_0 \eta}{\Delta P} \tag{2-19}$$

where M is a numerical factor usually between 0.5 and 1.5 which must be determined for individual cases, L is the total length of fibers per unit volume, and ϕ is defined by the equation:

$$\phi = \left[\ln \frac{1}{\beta} + 2\beta - \frac{\beta^2}{2} - \frac{3}{2} \right]^{-1} \tag{2-20}$$

M is a function of the degree of dispersion and the orientation of the fibers.

Langmuir's equations predict the existence of a particle size for minimum collection efficiency (minimum N/N_0).

Among the attempts to include the effect of inertia on γ is one by Davies.[19] Defining x as the distance from the axis of the fiber beyond which particles are not collected, he obtained

$$\gamma = \left(\frac{\beta}{1-\beta} \cdot \frac{2}{\pi R} \cdot \frac{x}{R} \right) t. \qquad (2\text{-}21)$$

This equation also predicts a particle size for minimum collection efficiency, but for smaller particle size values than those predicted by the Langmuir equation.

Dorman[2] developed a semiempirical equation based in part on dimensional analysis and applied it to data of Ramskill and Anderson:[20]

$$\log p = 2 - A'vt - D''tv^{-1/2} - It \qquad (2\text{-}22)$$

where p is the percentage penetration and A', D'', and I are inertial, diffusion, and interception parameters, respectively.

Dorman concluded that there is reasonable agreement between the values predicted by the Langmuir equation and the experimental results of Ramskill and Anderson for the interception parameter when a value of $M = 0.5$ is chosen to calculate the mean fiber radius (equation 2-19). However, then the diffusion parameter estimated from the experimental values does not agree with that derived from Langmuir's equations. He also concluded that at relatively high velocities his inertial parameter becomes important and depends on v^2.

Numerous other experimental studies have been made of fiber filter efficiencies. For example, Stern, Zeller, and Schekman[21] determined the

Table 2-1. Physical Properties of the Institute of Paper Chemistry (IPC) Fibrous Filter Mats.[21]

Material	Viscose
Mat thickness	0.084 cm
Average fiber diameter	17 μm
Fiber density	1.49 g/cm^3
Bulk density	0.189 g/cm^3
Average pore diameter	170 μm $\pm$ 60
Porosity	90%
Fiber volume	10%

collection efficiency of an Institute of Paper Chemistry fiber filter for monodispersed aerosols of 0.026–1.71 μm diameter at reduced pressures to simulate air densities in the stratosphere. These filters have been used extensively for stratospheric monitoring and research.[22] Filter efficiencies were determined at the ambient pressure in the laboratory and at pressures of 465, 140, 53, and 24 millibars. The physical properties of these filters are shown in Table 2-1 and their collection efficiencies at 53 millibars (40 torr) are shown in Figure 2-6. They reached the following conclusions:

1. A minimum in the collection efficiency-flow rate curves exists for each particle size and each pressure.
2. A unique velocity exists for which the collection efficiency is the same for all particle sizes. This velocity decreases with decreasing pressure.
3. For a given particle size and face velocity, the efficiency increases with decreasing pressure.
4. In the diffusion regime there is no particle size of maximum penetration, contrary to Langmuir's theories.
5. In both the impaction and diffusion regimes the collection efficiency curve for a dioctylphthalate (DOP) liquid droplet aerosol of diameter 0.3 μm was nearly the same as that for a polystyrene aerosol. Thus the IPC filter does not discriminate between liquid and solid particles. How-

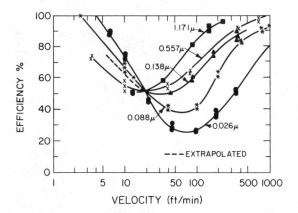

Figure 2-6. Experimental collection efficiencies of the IPC filter mat for spherical polystyrene particles of density 1.05 and spherical poliomyelitic particles of density 1.23 at 53 mb pressure.[21]

ever, this type of filter is impregnated with an oil to improve the retention of solid particles.

Stern et al.[21] compared their results with the equations for single fibers, using the following equation to estimate the efficiency of a single fiber, E_s, from the overall filter efficiency, E_0:

$$E_0 = 1 - \exp\left[-\frac{4}{\pi}\frac{\alpha t E_s}{D}\right] \tag{2-23}$$

where α = bulk density of the filter/density of the fiber, and t is the thickness of the filter measured parallel to the direction of flow. They compared their data for the diffusion regime with the predictions from Langmuir's equation for that regime (2-12) and from an equation by Torgeson:

$$E_s = 0.755 \left(\frac{C_{Da}\mathrm{Re}}{2}\right)^{0.4} \mathrm{Pe}^{-0.6} \tag{2-24}$$

where C_{Da} is the drag coefficient for a fiber imbedded in a mat. The dimensionless factor $(C_{Da}\mathrm{Re}/2)$ can be calculated from the equation

$$\frac{C_{Da}\mathrm{Re}}{2} = \frac{\pi}{4}\frac{\Delta p}{v}\frac{1-\alpha}{\alpha}\frac{D^2}{\eta t} \tag{2-25}$$

Stern et al.[21] obtained quite good agreement between their experimental data and Torgeson's diffusion theory. They pointed out that the Langmuir theory yields values for E_s that are approximately one-half of those predicted by Torgeson's theory.

Stern et al.[21] found that in the impaction regime the single fiber efficiency was nearly equal to the classical impaction parameter K, defined by equation (2-8). They also compared their results for this regime with predictions by an empirical equation of Davies[19] for E_s, for low Reynold's numbers, and by combining the mechanisms of diffusion, interception, and impaction they found

$$E_s = (0.16 + 10.9\alpha - 17\alpha^2)\cdot[R + (0.5 + 0.8(d/D))(K + D')$$
$$- 0.1052(d/D)(K + D')^2] \tag{2-26}$$

The diameter of the fiber to be used in equation (2-26) was calculated from another empirical equation by Davies:

$$\Delta p = \frac{70\eta v t \alpha^{1.5}}{D^2}(1 + 52\alpha^{1.5}) \tag{2-27}$$

The agreement between the experimental results and (2-26) was "satisfactory." Stern et al.[21] point out that equation (2-26) was not verified by other investigators who used glass fiber filters, and they suggest that "bounce" off of the fibers may have been responsible.

More recent comparisons of the theories of filtration efficiency and experimental results have been made by Friedlander[23] and by Spielman and Goren.[24] Friedlander extended the theory of diffusional exchange between flowing fluids and cylinders to the case of diffusing particles of finite diameter. He compared his theoretical predictions with filtration data obtained by Chen[1] and by Wong, Ranz and Johnstone,[15] concluding that there was good agreement. Spielman and Goren were primarily concerned with pressure drop but discussed the implications of their model for collection efficiency and compared selected data with the theoretical predictions of particle collection by interception and diffusion, interpolating through the region of combined effects. Impaction was assumed to be negligible. Again, good agreement was obtained.

The flow of a gas through a filter is accompanied by a pressure drop between the faces of the filter. Important as collection efficiency is in evaluating the performance of a filter, the pressure drop is often of equal or even greater importance. The pressure drop may control the flow rate, and this effect may become critical when sampling a very dilute aerosol. Also, long sampling times are required to collect sufficient particulate matter for measurement. A comparison of filter performance on the basis of both pressure drop and collection efficiency, which has been used by Cadle and Thuman,[25] makes use of the parameter F, defined by the equation:

$$\frac{N_1}{N_0} = \exp \frac{-F\Delta p}{100} \tag{2-28}$$

where N_0 and N_1 are, as usual, the number concentrations of particles entering and leaving the filter, respectively, and Δp is the pressure drop. Dorman[3] also mentions the use of this parameter. To make meaningful comparisons F must be determined using the same flow rates and particle sizes.

Numerous equations relating pressure drop to the flow rate, packing density, and fiber thickness have been proposed. One of these by Davies[5] (equation 2-27) has already been mentioned, and was developed empirically for filters having a high packing density.

Chen[1] pointed out that two approaches have been used to develop equations predicting the pressure drop across filters. One, called channel theory, considers porous beds to consist of a system of interconnected channels. The pressure drop through them for viscous flow is based on D'Arcy's equation:

$$\Delta p = k_0 v \eta t \tag{2-29}$$

where k_0 is a constant. Later k_0 was theoretically evaluated to yield the Kozeny–Carman equation:

$$\Delta p = \frac{k_1 v \eta S_0{}^2 (1 - \epsilon)^2 t}{\epsilon^3} \tag{2-30}$$

where S_0 is the specific surface of the packing material, ϵ is the porosity of the filter ($=1 - \alpha$), and k_1 is a constant: The Kozeny–Carman equation has been used extensively to determine the specific surfaces and mean particle sizes of powders. Chen refers to an equation by Sullivan and Hertel[26] who reviewed this approach to the problem:

$$\Delta p = \frac{k_2 \eta v t S_0{}^2 (1 - \epsilon)^2}{k_3 \epsilon^3} \tag{2-31}$$

where k_2 is a shape factor and k_3 an orientation factor.

The other approach is called drag theory and grew out of the finding by Brinkman[27] that the equations based on the channel theory do not apply to very porous media such as most fiber filters. Spielman and Goren[24] applied Brinkman's model for flow through very porous media to flow through fibers to predict the pressure drop through fiber filters having several different arrangements of the fibers. They assumed that the fluid near a fiber embedded in a filter experiences, in addition to the usual force terms, a "body-damping" force that is proportional to the local velocity and which accounts for the influence of neighboring objects on the flow. Some of their predictions are shown in Figures 2-7 and 2-8.

Chen[1] compared experimental results for the pressure drop across filters with a number of the equations for pressure drop. He states that for the most part the results only demonstrate the proportionality to the superficial velocity and thickness of the filter and to the empirical exponential function of porosity.

2.1.2. Membrane Filters

The filtration process by membrane filters is not particularly well understood. Of the two main types, Millipore and Nuclepore filters, the structure of the former is more complicated than that of the latter (Figures 2-2 and 2-3). The Millipore filters have three separate structures, the upper surface, interior, and lower surface, whereas the Nuclepore filters

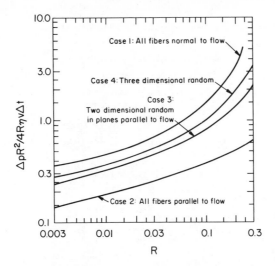

Figure 2-7. Theoretical predictions of dimensionless pressure gradient versus volume fraction fibers for different fiber orientations for no slip. Reprinted with permission from Spielman, L., and S. L. Goren, *Environ. Sci. Tech.* **2,** 279 (1968). Copyright by the American Chemical Society.

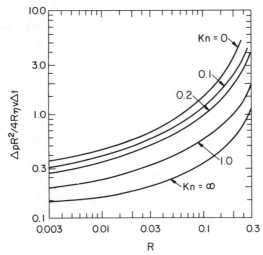

Figure 2-8. Theoretical predictions of dimensionless pressure gradient versus volume fraction fibers for different Knudsen numbers where fiber axes are all perpendicular to the direction of flow. Reprinted with permission from Spielman, L., and S. L. Goren, *Environ. Sci. Tech.* **2,** 279 (1968). Copyright by the American Chemical Society.

61

have holes that are circular in cross section and have a smooth upper surface.

The theory of aerosol filtration by Nuclepore filters has been investigated by Spurny et al.[28] To develop an equation for the collection efficiency they used equations for the partial efficiencies due to impaction, diffusion, and interception, and then combined the partial efficiencies using an empirically evaluated weighting factor.

The partial efficiency due to impaction, E_I, used by Spurny et al.[28] was

$$E_I = \frac{2E_I'}{1 + z} - \frac{2E_I'^2}{(1 + z)^2} \tag{2-32}$$

where

$$E_I' = 2K\sqrt{z} + 2K^2 z \exp\left(-1/K\sqrt{z}\right) - 2K^2 z \tag{2-33}$$

$$z = \frac{\sqrt{P}}{1 - \sqrt{P}} \tag{2-34}$$

P = pore volume/filter volume, K is the inertial parameter (here expressed as $mv/6\eta r R_0$), m = mass of a single aerosol particle, and R_0 is the pore radius of a clean filter.

These authors also developed semiempirical equations to estimate the change in pore size resulting from the collection of droplets or solid particles as a function of time. For droplets the equation for the pore diameter (R_T) at time T seconds is

$$R_T = \left[R_0^2 - 1.33r^3 N_0 T \frac{v}{\lambda N_p}\right]^{1/2} \tag{2-35}$$

where r = aerosol particle radius in centimeters, N_0 is the aerosol particle concentration in number per cubic centimeter, and N_p is the number of pores per square centimeter of filter surface. For solid particles the equation is

$$R_T = \left[R_0^2 - 2\left(\frac{R_0}{(R_0 + ar)}\right) r^3 N_0 T \frac{v}{\lambda N_p}\right] \tag{2-36}$$

where a is the "coefficient of compactivity" and is given by

$$a = \left(\frac{R_0}{r}\right)^{0.57} \tag{2-37}$$

Equation (2-32) can be corrected for the changing efficiency with time by substituting R_T from equations (2-35) or (2-36) for R_0 in the expression for the inertial parameter, K.

Spurny et al.[28] used two different equations to calculate the partial efficiency of diffusion, the choice of equation depending on the value of a parameter N_D $(=tD'P/R_0{}^2v)$, where D' is the diffusion coefficient for the particles in square centimeters per second. For the range $N_D < 0.01$ they used the equation

$$E_D = 2.56 N_D{}^{2/3} - 1.2 N_D - 0.177 N_D{}^{4/3} \qquad (2\text{-}38)$$

and for $N_D > 0.01$

$$E_D = 1 - 0.81904 \exp(-3.6568 N_D) - 0.09752 \exp(-22.3045 N_D)$$
$$- 0.03248 \exp(-56.95 N_D) - 0.0157 \exp(-107.6 N_D)$$
$$- \cdots \qquad (2\text{-}39)$$

Again, R_T could be substituted for R_0.

The partial efficiency of interception, E_R, was obtained from the expression

$$E_R = N_R(2 - N_R) \qquad (2\text{-}40)$$

where

$$N_R = r/R_0 \qquad (2\text{-}41)$$

The formula is invalid when $N_R > 1$. And again, R_T can be substituted for R_0. Spurny et al.[28] combined the partial efficiencies with the equation

$$E = E_I + E_D + \delta E_R - E_I \cdot E_D - \delta E_I \cdot E_R \qquad (2\text{-}42)$$

The value of δ was 0.15, and this was interpreted as meaning that the contribution of interception was only 15% of that given by theory.

The authors also treated the pressure drop across Nuclepore filters, considering three regimes of gas flow through the filters. Defining the Knudsen number now in terms of the pore radius ($Kn = \lambda/R_0$), when $Kn \ll 1$ the pressure drop was represented by the Hagen–Poiseuille equation in the form:

$$\Delta p = p - p \left(1 - 5.093 \frac{\eta t}{p} \cdot \frac{v}{R_0{}^4 N_p} \right)^{1/2} \qquad (2\text{-}43)$$

For somewhat larger values of $Kn (<1)$ a correction was introduced for gas slippage:

$$\Delta p = p - p \left[1 - 5.093 \frac{\eta t}{p} \left[\frac{v}{R_0{}^4 N_p} \left(\frac{1 + 5.50r}{R_0} \right) \right] \right]^{1/2} \qquad (2\text{-}44)$$

When Kn was near unity or larger, they used the equation

$$\Delta p = p + 3.42A'\eta/R_0 \left(\frac{2\pi R_g T}{M}\right)^{1/2}$$

$$- \left[\left[p + \frac{3.42A'\eta}{R_0}\left(\frac{2\pi R_g T}{M}\right)^{1/2}\right]^2 \right.$$

$$\left. - 5.093(\eta + p)\frac{v}{R_0^4 N_p}\right]^{1/2} \tag{2-45}$$

where R_g is the gas constant, 8.313×10^7 gm cm^2 sec^{-2} deg^{-1}; T is the absolute temperature, $^\circ K$; M is molecular weight of the gas; A' is a constant, here taken as 0.75. Once again, R_T can be substituted for R_0. All the equations in the Spurny et al.[28] paper had dimensions in the cgs system.

These authors measured E and Δp with several different aerosols. They concluded that their filtration theory accurately predicts the behavior of nuclepore filters in all respects except time dependency, and even here the theory gives good qualitative predictions.

Figure 2-9 is an electron micrograph of polystyrene particles collected on a nuclepore filter.

The theory of millipore filters has been investigated by Pich[4,29] and by Spurny and Pich.[30] For impaction Pich used equation (2-32), which he had originally derived. He used equation (2-39) for the parameter $N_D' = D't/R_0^2 v > 0.03$, and for that parameter < 0.03, the equation

$$E_D = 2.57(N_D')^{2/3} \tag{2-46}$$

or equation (2-38). He also discussed equations for gravitational deposition in the pores, the sieve effect (when particles are greater than the pore diameter), and interception when the particles are smaller than the pore diameters. The total efficiency of the millipore filters was expressed by the equation

$$E = E_I + E_D - E_I E_D \tag{2-47}$$

Thus it is evident that the theoretical work of Spurny et al.[28] was a considerable elaboration of the earlier work of Pich for a very similar type of filter.

Both these theories for membrane filters predict minima in the efficiency versus particle size curves, and this prediction has been borne out by several of the experimental studies.[4,28,29] For example, Figure 2-10 is a comparison by Spurny et al.[28] of the computed experimental dependency of collection on particle size. The membrane filters behave somewhat like fiber filters in this respect, because physically the branches of

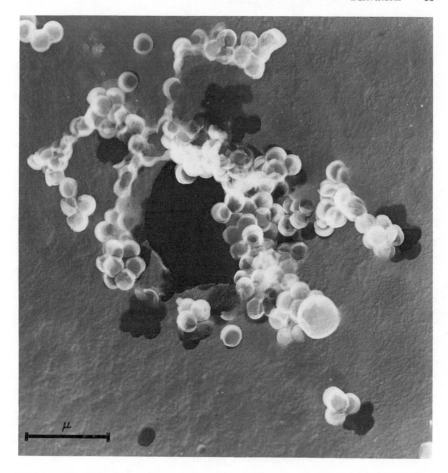

Figure 2-9. Electron micrograph of polystyrene particles. Magnification 21,180×.

the curve represent different mechanisms of collection, the branch for smaller particles representing largely collection by diffusion. Other workers, for example, Megaw and Wiffen[31] and Walkenhorst[32] found no minimum, but different workers have used different makes of filters and somewhat different conditions, so it is difficult to compare their results.

2.2. Characteristics of Various Filters for Sampling Airborne Particles

Numerous representatives of the classes of filters discussed are available, or can be prepared, for collecting airborne particles prior to measure-

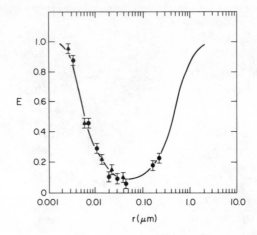

Figure 2-10. Comparison of computed (line) and experimental (points) dependency of efficiency on particle size. $R_0 = 2.5$ μm; $v = 5$ cm/sec; ● Selenium aerosol; △ pyrophosphoric acid aerosol. Reprinted with permission from Spurny et al., *Environ. Sci. Tech.* **3**, 453 (1969). Copyright by the American Chemical Society.

ment. The choice of a filter for a particular sampling operation depends on a number of factors such as the nature of the aerosol (concentration, size distribution, etc.), the method of measurement of the collected particles, and the collection conditions (e.g., laboratory sampling, stack sampling, sampling from aircraft). Table 2-2 shows some of the structural properties of commonly used filter media.

2.2.1. Fiber Filters

The main types of fiber filters used for aerosol sampling are those consisting of synthetic polymers, cellulose fibers, glass fibers, and mixed fibers. Detailed tables of the characteristics and commercial sources of many of these and other filter types are given in reference 33.

The synthetic polymer fiber filters have the advantages that they can be prepared in extremely pure form, they can be ashed, their manufacture is relatively simple, and they can be prepared with mean fiber diameters varying far down into the submicron size. They are commercially available under the name Delbag Microsorban (polystyrene fibers).

A method for preparing such filters on a small scale in the laboratory has been described by Cadle and Thuman.[25] A jet of polymeric material

Figure 2-11.   Handling polystyrene fiber filters in a "clean room" at the National Center for Atmospheric Research.

in solution in an organic solvent is introduced from a nozzle into a high-velocity air stream, is broken up, and is drawn into fibers from which the solvent evaporates. The spray nozzle is constructed so that an annular blast of air surrounds the jet of solution. The fibers produced are drawn by suction against Dacron netting until a mat of the desired thickness has accumulated. Extremely clean filters can be prepared by redistilling the monomer and polymerizing it with care to prevent contamination if all the operations are carried out in a clean room. Figure 2-11 demonstrates the handling of such filters in a clean room. Figure 2-12 shows some of the size distributions that can be obtained. Numerous combinations of solvent and polymer can be used. Fibers have been prepared from cellulose acetate, cellulose nitrate, polymethylmethacrylate, polystyrene, polyvinylchloride, and polyvinylformal. The combination polystyrene-methylenechloride produces especially good fibers. The concentrations of polymer in the solutions must be rather large (a few percent) to produce fibers instead of droplets.

Fibers for filter preparation have also been produced commercially and in the laboratory by the extrusion of thermoplastic materials.

Because of the purity of these filters and the fact that they can be ashed or dissolved in solvents, these filters are especially useful for de-

Table 2-2. Structural Properties of Nuclepore, Membrane, and Fiber

	1 $\overline{R}$ μm	2 $\overline{R}_m$ μm	3 $\sigma_g(\overline{R}_m)$	4 $\sigma(\overline{R}_m)$ μm	5 N_P per Sq Cm or Cm^{-2}	6 A_{PF}
Nuclepore						
	0.25	0.210	1.18	±0.08	3.10^7	5.10×10^{-2}
	0.40	0.241	1.07	±0.06	1.10^7	2.52×10^{-2}
	0.50	0.441	1.15	±0.12	1.10^7	3.90×10^{-2}
	1.00	0.862	1.13	±0.19	2.10^6	6.28×10^{-2}
	2.50	2.051	1.03	±0.14	4.10^5	7.80×10^{-2}
	4.00	3.402	1.08	±0.78	1.10^5	5.01×10^{-2}
Other Membranes						
VUFS-Synthesia	0.15	0.15		±0.02	10^6-10^8	10^{-1}
HUFS-Synthesia	0.40	0.23		±0.06	10^6-10^8	10^{-1}
HA-Millipore	0.23			—	10^6-10^8	10^{-1}
AA-Millipore	0.40			—	10^6-10^8	10^{-1}
Acropor-Gelman	0.40			—	10^6-10^8	10^{-1}
OH-Millipore	0.75			—	10^6-10^8	10^{-1}
SS-Millipore	1.50			—	10^6-10^8	10^{-1}
AUFS-Synthesia	0.7	1.55		±0.18	10^6-10^8	10^{-1}
RUFS-Synthesia	1.2	2.70		±0.31	10^6-10^8	10^{-1}
PUFS-Synthesia	3.5	3.10		±0.42	10^6-10^8	10^{-1}
OS-Millipore	5.00			—	10^6-10^8	10^{-1}
Fiber						
AGF-Gelman						
PF-41-Whatman						

[a] $\overline{R}$ = nominal pore radius
$\overline{R}_m$ = mean pore radius, measured by electron microscopy
σ = standard deviation. σ_g = geometry standard deviation
N_P = pore number per 1 sq cm
A_{PF} = pore area per 1 sq cm (equals porosity for Nuclepore)
Δp_1 = pressure drop for air velocity q = 1 cm/sec, filter area 6.18 sq. cm, ± mean deviation of 10 samples

Filters.[a] From Spurny et al.[28]

7	8	9	10	11	12
Δp_1 Mm H$_2$O	$(\Delta p)_A$ Mm H$_2$O/ Sq. Mm	$\overline{m}$ Mg	Δ_{max} Mg	$\overline{m}_n$ Mg/Sq Cm	L μm
222.3 ± 13.1	7.10	21.24 ± 0.03	±0.05	1.23	12
208.1 ± 9.7	3.38	27.70 ± 0.20	±0.40	1.58	11
30.6 ± 2.3	1.20	18.39 ± 0.08	±0.18	1.05	10
19.4 ± 2.1	0.51	18.10 ± 0.07	±0.08	1.10	8
11.1 ± 1.3	0.23	23.03 ± 0.22	±0.32	1.33	8
14.2 ± 1.6	0.45	20.20 ± 0.06	±0.12	1.16	10
342.1 ± 57.3	5.45	59.6*	±0.23	6.22	50–300
134.6 ± 5.1	—	35.6*	±0.23	4.76	50–300
135.4 ± 6.7	—	90.2	±0.23	5.19	50–300
57.5 ± 4.1	—	76.8	±0.23	4.45	50–300
268.3 ± 17.6	—	148.9	±0.23	8.59	50–300
253.1 ± 18.4	—	86.4	±0.23	5.00	50–300
23.2 ± 2.1	—	60.2	±0.23	3.48	50–300
6.5 ± 1.1	—	19.8*	±0.23	2.07	50–300
4.8 ± 2.3	—	37.5*	±0.23	3.91	50–300
4.7 ± 0.7	—	27.7*	±0.23	2.89	50–300
6.3 ± 0.9	—	123.8	±0.23	7.15	50–300
14.7 ± 0.8	—	147.7	±0.27	8.53	200–500
10.8 ± 1.6	—	81.6	±0.27	8.51	200–500

$(\Delta p)_A$ = pressure drop per 1 sq mm of pore area

$\overline{m}$ = mean weight of filter, 4.7-cm diameter unless otherwise indicated, ± mean deviation of 10 samples

Δ_{max} = maximum deviation from m, 10 samples

$\overline{m}_n$ = mean weight per 1 sq cm of filter

L = filter thickness

* filter diameter 3.5 cm

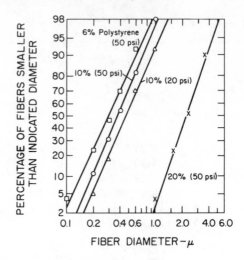

Figure 2-12. Diameter distributions of fibers obtained from solutions of poly-styrene in methylene chloride of various concentrations and air pressures. Re-printed with permission from Cadle, R. D., and W. C. Thuman, *Ind. Eng. Chem.* **52,** 316 (1960). Copyright by the American Chemical Society.

termining the chemical or elemental composition of airborne particles, for example, by neutron activation.

Composite filters have been used extensively, for example, by the U.S. Chemical Warfare Service. The usual practice is to combine asbestos fibers, which are quite fine, with a coarser fiber to prevent matting.[34] The coarser material has included wool,[35] cotton,[36] and cellulose.[37] Glass-asbestos and cellulose-glass mixtures have also been used. Remov-ing the collected particles from such filters is very difficult; they are mainly useful for sampling when only the weight of the collected mate-rial is to be measured, or when radioactive particles are to be collected and the radioactivity measured directly on the filters.

Glass fiber filters are available in a number of forms, with and with-out organic binders. They have the advantages of being able to with-stand fairly high temperatures and of being nonhygroscopic. Since they are not soluble in organic solvents except when a binder or lubricant is present, the material collected on them can be analyzed for soluble organic material by extraction with an appropriate solvent. Ceramic fibers have been used for temperatures up to 1800°C.[38,39]

Cellulose fiber filters of the types used by chemists for filtering aqueous suspensions have been used extensively for collecting aerosol particles. They are available in many different grades, including some with a very low ash content. Among their disadvantages are that their collection efficiencies are smaller than those of many other filters, they are nonuniform, and they are quite hygroscopic. Whatman No. 41 filter is perhaps the brand and grade most widely used for aerosol sampling. It is a very pure paper, with rather low flow resistance. Some such papers have been hardened and are useful when the particles are to be removed, for example, by supersonic agitation. Various cellulose filter papers are available for special purposes, such as "high-volume" sampling, described below.

Stainless steel wool filters have also been used occasionally, especially for high temperatures and corrosive aerosols.

2.2.2. Membrane Filters

An advantage of membrane filters mentioned earlier was that most of the collected particles and all of those larger than the holes are collected on the surface of the filter sheet. Other advantages are that they become transparent when a drop of liquid is added which has about the same refractive index as the filter material, and that they are soluble in many organic liquids. Both of these properties can be used to advantage when examination of the collected particles is contemplated. Disadvantages are the relatively large flow resistance and the rather brittle character of the membranes.

The two main types are those in which the pores are introduced when the sheets are formed, including the millipore filters, and those in which the pores are formed later, especially the nuclepore filters. The microstructure of these two types is shown in Figures 2-2, 2-3, and 2-9. The former type is manufactured by many companies in different parts of the world and of a number of different polymeric materials including cellulose triacetate, mixtures of cellulose, polyvinylchloride, and acrylonitrile. The pore diameters generally cover the range of 10–0.01 μm. However, as discussed earlier, particles much smaller than the pore sizes can be collected efficiently, evidently because they are collected by diffusion.

Nuclepore filters are manufactured from 10-μm-thick sheets of "polycarbonate" which are placed in contact with uranium sheets and put into a nuclear reactor. The uranium-235 in the uranium sheets undergoes fission; the fission fragments produce tracks of chemical reaction products in the plastic sheets. These products are removed in an etching bath, where the pores are also enlarged to an extent determined by the type of reagents in the bath, the bath temperature, and the duration of etching.

Nuclepore filters, manufactured by the General Electric Company, are available with fairly uniform pore sizes ranging from 0.5 to 8.0 μm in diameter. Unlike the other membrane filters, the pores are circular in cross section, and the filters are quite strong.

2.2.3. *Miscellaneous Filters*

Granular filters of various types, such as sintered disks, have sometimes been used. They are sometimes prepared from sugar, salicylic acid, or naphthalene. When sampling is completed the filter material is dissolved or evaporated.

Many of the filter types described above are available in a variety of forms. Some are available in the form of thimbles and porous paper bags, which have been used for stack gas sampling. Some have been pleated, which increases the available filter area for a given superficial area. These are used extensively in the high-volume filters.

Resin-impregnated filters are sometimes used, which apparently owe their high collection efficiencies to the fact that the resin particles have a low conductivity and thus retain electrical charges for a long time period. These filters, however, are seldom if ever used for aerosol characterization, at present being used largely in respirators. According to Davies[5] the filters are negatively charged by contact or friction.

2.3. Filtration Accessories

A great array of pumps, filter holders, and filter changers is available or has been described in the literature. Criteria to be used in the selection of a pump include the sampling rate desired, the pressure drop across the filter, and the power requirements. When the filter resistance to flow is small and the desired air flow rate is large, a simple fan-type pump is often satisfactory. This is the situation often encountered when monitoring for particulate matter in the contaminated air of cities or in and around industrial plants. The so-called high- volume samplers are often used for such purposes. They originally consisted of a vacuum cleaner motor and pump connected to a conical filter holder which held a pleated filter. They are now commercially available in a number of modifications including portable samplers that combine the motor, pump, and filter holder in a simple unit equipped with a handle. High-volume samplers are generally extremely noisy, which makes them impractical for many purposes. Such samplers generally require a source of line power because of the high wattage requirements. However, a number of battery-operated air filter samplers are commercially available for use where lower flow rates are acceptable.

When the resistance of the filter to air flow is very large, an oil-soaked vane vacuum pump may be required to provide the pressure drop across the filter needed to obtain a reasonable flow rate.

In some circumstances no pump at all is needed. An example is the main filter sampling system carried on the U.S. Air Force RB-57F aircraft and originally designed to collect samples of particles from the atmosphere to determine atmospheric radioactivity levels. In this case the flow of the airstream past the aircraft forces the air through the filter. When the filters are not deployed they are held in twelve filter holders within what would normally be the bomb bay of the aircraft. By means of a switch in the navigator's compartment, they can be lowered sequentially with a "record changer" mechanism into a sampling probe mounted beneath the aircraft. Figure 2-11 shows one of these filter holders.

Whenever possible, open-faced filter holders should be used since they avoid the errors resulting from the use of probes and of tubing to conduct the air being sampled. These problems and ways of minimzing them are discussed in a later chapter. Backing, consisting of a porous material, a mesh, or a screen for the filter, is usually required; for quantitative work a seal around the edges of the filter to prevent the aerosol bypassing it is essential. Figure 2-13 is a schematic drawing of an openface filter holder which holds eight filters that can be rotated sequentially into sampling position. It has been used at the National Center for Atmospheric Research for sampling in the field with Nuclepore filters.[39]

For quantitative work the flow rate through the filter must be known. Numerous types of flowmeters are available, but often it is convenient to maintain a constant, predetermined flow rate. This can be achieved with a critical orifice, which can be used with many types of air sampling equipment in addition to filters. A critical orifice is merely a constriction in the air-flow line, preferably between the filter and the pump, across which the pressure drop exceeds $\frac{1}{2}$ atm. The flow rate through the orifice remains relatively constant despite changes in the air resistance in the sampling apparatus. A hypodermic syringe needle is a convenient critical orifice when the desired flow rate is small.[40] It can be used in a sampling system such as that shown in Figure 2-14 and used extensively by the author. The vacuum manifold is made of plastic or rubber tubing and the needles are inserted through the wall. The system shown in the figure includes bubblers for trace gas sampling and plastic tubes containing

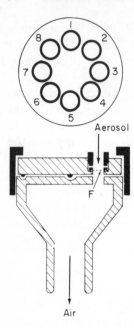

Figure 2-13.   Schematic drawing of an open-face filter holder. Reprinted with permission from Spurny *et al.*, *Environ. Sci. Tech.* **3,** 464 (1969). Copyright by the American Chemical Society.

silica gel to remove spray from the liquid reagents in the bubblers which might otherwise clog the needles. The bubblers and protective tubes could be omitted when only filtering is desired. Experience has shown that the flow rate through the needles begins to decrease after a few days; replacement of the needles each day is good practice.

Other types of flow-rate regulators have been used. For example, the Giraffe Constant Flow Air Sampler (Flow Master Products, Inc.) uses a Gast 0521 pump and has a flow-regulating valve between the pump and the filter. The valve is essentially a variable orifice that opens and closes to compensate for changes in the resistance of the filter.

Various automatic sequential samplers that can be used with filters are available commercially. For example, the Gelman Sequential Sampler has been designed primarily for use with filter holders or midget impingers (described later). It will collect up to twelve samples in timed sequence, and once set in operation it will operate unattended and will shut itself off automatically. The most important part of such an instrument is the sequential valve which, for the Gelman instrument, is controlled by an adjustable timer that automatically switches the vacuum pump in turn to each of the intake ports. The IDC Sequential Sampler (Instrument Development Co.) is another such instrument, and is de-

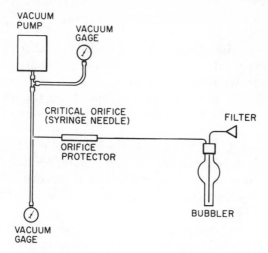

Figure 2-14. Schematic of aerosol (and gas) sampling apparatus.

signed to collect twelve samples over a 12-hour period. The length of the
sampling time can be set to any value between zero and 30 minutes.
During the remainder of each hour the pump is turned off until a new
sampling head is switched in.

Paper-tape samplers are also available from a number of companies
and are generally used to obtain a record of the concentrations of particulate air pollution. The tape consists of filter material, and the samplers automatically collect samples by drawing air through the tape with
a vacuum pump. The tape is moved at preset intervals, and the length
of time sampled on each portion of tape is controlled by adjustable
timing devices.

3. SEDIMENTATION

One of the simplest and oldest methods for collecting particles, both
liquids and solids, is by sedimentation. It was used, along with other
methods, in the pioneering work by Whytlaw-Gray[41] who apparently
coined the term "aerosol" to mean a disperse system of particles suspended in a gas. It has been largely replaced as a sampling method in
recent years by more sophisticated and dependable methods, but it is
still occasionally used because the equipment is readily built, is inexpensive, is simple to undertake, and for some purposes is the best
method, even when more elaborate techniques are available.

3.1. Theory

Particles suspended in a fluid of density lower than that of the particles settle through the fluid; if the particles are sufficiently far apart the settling rate increases until a constant rate, known as the terminal velocity, is obtained. The terminal velocity depends on the size, shape, and density of the particles and on the viscosity and density of the fluid. If the particles are small relative to the mean free path of the molecules or atoms of the fluid (i.e., if the Knudsen number, Kn, is large), the terminal velocity also depends on the mean free path, because the fluid cannot be considered to be a continuum. Under these conditions "slippage" occurs, and the particles undergo Brownian motion and diffusion which is superimposed on the sedimentation.

Stokes derived the basic equation for the motion of a spherical particle suspended in a fluid and subjected to a constant force, f. That force can be equated to the opposite viscous force exerted by the fluid when the particle attains a steady velocity v:

$$f = 6\pi r \eta v \tag{2-48}$$

The downward force on a particle falling in a fluid under the influence of gravity is

$$f = \frac{4}{3} \pi r^3 (\rho - \rho') g \tag{2-49}$$

where, as usual, ρ is the particle density and ρ' is the density of the fluid. The opposing viscous force increases with the increasing rate of fall of the particle until the terminal velocity is attained. Then

$$\frac{4}{3} \pi r^3 (\rho - \rho') g = 6\pi r \eta v \tag{2-50}$$

and

$$v = \frac{2gr^2(\rho - \rho')}{9\eta} \tag{2-51}$$

This is the well-known Stokes equation; it assumes that the fluid is continuous and that viscous drag is the only restraining force on the particle. These assumptions are invalid when the particles are sufficiently large that a turbulent wake forms or so small that the Knudsen number exceeds unity and slippage occurs. The latter situation produces a terminal velocity greater than that predicted by equation (2-51). Since the particle density usually greatly exceeds the gas density, equation (2-51) reduces to

$$v = \frac{2gr^2\rho}{9\eta} \tag{2-52}$$

Numerous comparisons have been made of the velocities predicted by equation (2-51) with experimentally determined rates of fall, and the agreement is excellent when the Reynolds number and the Knudsen number approach zero. Therefore, the Stokes relationship has been applied to the determination of particle size with great confidence.

A number of early investigators, especially Millikan,[42,43] Knudsen and Weber,[44] and Mattauch,[45] attacked the problem of correcting for slippage. The subject has been reviewed in detail by Fuchs.[16] For most purposes the so-called Cunningham correction is satisfactory:

$$f = \frac{6\pi r\,\eta v}{1 + A(\lambda/r)} \tag{2-53}$$

thus

$$v = \frac{2gr^2(\rho - \rho')}{9\eta}\left(1 + A\,\frac{\lambda}{r}\right) \tag{2-54}$$

The A is a constant close to unity (0.9 is often used) and varies somewhat depending on the particle material, since A depends on the fraction of gas molecules that collide with the particle that undergo specular reflection. Millikan[43] showed by means of kinetic theory that A varies from a lower theoretical limit of 0.7004 (diffuse reflection) for small values of Kn to 1.164 for large values of Kn. He therefore put $A = A' + B \exp(-Cr/\lambda)$. Using an oil-drop method he determined $A' = 0.864$, $B = 0.290$, and $C = 1.25$. He also found that although A' varies with the nature of the gas and even more with the material of the droplet, $(A + B)$ is constant within 2 or 3% for most particles that might settle through the atmosphere. Fuchs[16] has estimated that the error in using equation (2-53) for oil droplets in air at 23°C and 760 torr is 1% for $r = 0.18\ \mu$m and 10% for $r = 0.05\ \mu$m, whereas equation (2-48) gives an error of 1% for $r = 8\ \mu$m and 10% for $r = 0.8\ \mu$m.

Sinclair[13] states that for particles having radii between 2 and 0.1 μm, the settling velocity can be estimated from equation (2-52) by adding 0.04 μm to the radius.

As mentioned above, the Stokes equation was derived omitting the inertia terms from the equation of motion, that is, by assuming a very small value for the Reynolds number, Re, here defined as $2rv\rho'/\eta$.

Fuchs[16] states that the error in the Stokes equation is approximately proportional to Re and is about 1.7% for Re = 0.1. Davies[46] has estimated that errors of 1, 5, and 10% are produced by the Stokes equation for Reynolds numbers of 0.074, 0.38, and 0.82, respectively. From equation (2-52) and the definition of the Reynolds number the following approximate equation can be obtained relating the maximum particle size, r_c, that should be used to avoid exceeding a given error to the maximum Reynolds number:

$$r_c{}^3 = \frac{9\eta^2 \text{Re}}{4g\rho\rho'} \qquad (2\text{-}55)$$

The viscosity of air is about 1.82×10^{-4} poise and the density about 1.2×10^{-3} g cm^{-3} at 760 torr and 20°C. Thus for particles of 2 g cm^{-3} density settling in air, the particle radii corresponding to Re = 0.074, 0.38, and 0.82 are about 13, 22, and 63 μm, respectively.

Several equations have been proposed to correct the Stokes equation for viscous drag. Davies[46] has suggested the use of the following equations:

$$\psi = \text{drag coefficient} = \frac{(m - m')\, g}{\pi(d^2/4) \cdot \tfrac{1}{2}\, \rho v^2} \qquad (2\text{-}56)$$

and

$$\text{Re} = \frac{\psi \text{Re}^2}{24} - 2.3363 \times 10^{-4}(\psi \text{Re}^2)^2 + 2.0154$$

$$\times\, 10^{-6}(\psi \text{Re}^2)^3 - 6.9105 \times 10^{-9}(\psi \text{Re}^2)^4 \qquad (2\text{-}57)$$

where m and m' are the mass of the spherical particle and of the fluid displaced by the particle, respectively. These equations yield accurate results when the Reynolds number is as high as 4.

Strictly speaking, particles must be rigid if they are to obey the Stokes equation, and most obviously solid particles fulfill this requirement. Also, for drops of high viscosity falling through a medium of low viscosity, circulation effects are negligible throughout the particle size range where the Stokes equation applies.

The above equations were developed for small solid spheres and do not necessarily apply to particles of other shapes. As would be expected, numerous investigators have studied the effect of particle shape on the settling velocity. Whytlaw–Gray and Patterson[41] were among the earliest to make such investigations. They calculated that in the case of an ellipsoid having a ratio of axes of 2:1, or even 3:1, the average value of r to be used in the Stokes equation is nearly the same as the radius of the sphere having the same volume. Fuchs[16] has discussed the question of the resistance of a medium to the motion of nonspherical particles in considerable detail.

The above discussion is based on the assumption that the fluid is at rest. When this is the situation, except for secondary effects (agglomeration, etc.), the concentration at any given level in the suspension of particles of a given size remains constant until all particles of that size have settled below that level. This is so because, on the average, for every particle that settles out of that level, one settles in to take its place.

The situation is entirely different when the settling suspension is stirred. A random motion is then superimposed on the downward motion of the particles. Since the upward motions in a closed system such as a box are compensated by downward motions, the stirring does not affect the average rate of fall, but rather, tends to maintain a uniform concentration throughout the suspension that gradually decreases with time. There is also a correponding decrease in the rate of settling with time.

The basic equation for such settling in a monodispersed aerosol in a box is

$$\frac{v}{h} \, dt = - \, \frac{dN}{N} \tag{2-58}$$

where h is the height of the box. In integrated form, equation (2-58) is

$$\frac{N}{N_0} = \exp - \frac{vt}{h} \tag{2-59}$$

When the suspension is polydispersed, the particles having the same diameters will settle as predicted by equations (2-58) and (2-59), except for secondary effects such as the larger, faster settling particles impacting on the smaller ones.

Sinclair and others[13,47] obtained the following equation for a polydispersed aerosol having a log-normal size distribution undergoing stirred settling:

$$- \frac{d}{dt} \log C_t = 1.3 \times 10^5 \, \frac{\rho}{h} \, d^2 \tag{2-60}$$

where d is the medium weight diameter and C_t is the cross-sectional area per cubic centimeter of the particles.

The above theory is useful for designing equipment for obtaining samples of aerosol particles by sedimentation, especially when the aerosol is captured in a closed container or is flowed through a duct such that the sedimenting particles are collected on the bottom. A difficulty in applying the theory is that quiet settling and stirred settling are limiting situations, and often there is turbulence without thorough mixing. Also, sedimentation techniques are often applied to the open atmosphere, in which case the interpretation of the results is quite different.

Schadt and Cadle,[48] as part of a study of the collection efficiencies of commonly used aerosol sampling devices, investigated the effect of the collection surface and particle material on the collection behavior of particles in a box. The sedimentation box was constructed of Styrofoam (for thermal insulation) and the interior was lined with aluminum to provide an electrically conducting surface, which was grounded. The top of the box was hinged so that it could be opened, filled with aerosol, and rapidly closed, and so that collecting slides could be readily inserted or removed. Microscope slides with and without electrically conducting layers of Aquadag (Acheson Colloids Co., Port Huron, Mich.) were used as collecting surfaces for subsequent microscopic examination. The inside dimensions of the box were 25.4 cm high and 35.6 × 40.6 cm cross section.

Monodisperse stearic acid aerosols were prepared with a Sinclair–LaMer condensation-type generator, and the particle sizes were determined using an "Owl" (described elsewhere in this book) and microscopic examination of the collected particles. Most of these aerosols exhibited higher order Tyndall spectra, indicating that they were nearly monodispersed. Aqueous solutions of sodium chloride were dispersed using an aspirator-type aerosol generator of the type described by Cadle and Magill.[49]

Concentrations of particles determined by counting those deposited on the slides were compared with those determined by filtering the aerosols with membrane filters and counting the particles collected on the filters. Stearic acid aerosols of 1- and 2-μm diameter produced very erratic, uneven distributions of deposit on the slides in the sedimentation box and consistent differences for the conducting and nonconducting slides. Concentrations of 2-μm particles determined by the sedimentation method were considerably higher than those determined from membrane filter counts. The agreement between the two methods was fairly good for larger particles.

Excellent agreement was obtained between the concentrations as obtained with the membrane filters and by sedimentation for 0.4-μm sodium chloride particles. Very even distributions of particles on the slides on which the particles settled was also observed. The differences in the sedimentation behavior of the two types of particle materials may have been a result of their different electrical properties, which may have affected the motions of the particles as they approached the surfaces of the box and of the slides.

These results suggest that sedimentation is a rather untrustworthy method for quantitatively collecting particles smaller than a few microns in diameter, even when the aerosol is in a closed container.

3.2. Applications

Numerous variations of the sedimentation technique have been used. Schaefer[50] has deployed "sedimentation foils" which are exposed for month-long periods. Strips of aluminum foil are placed in specially designed holders. These strips are coated with a sticky but very stable silicone adhesive that remains effective for more than 2 months. Schaefer states that using a microbalance the increase in weight due to deposited particles can be measured even for a single day's exposure in an area having light-to-moderate air pollution. The holder is designed to prevent contamination by birds. The system has been used in New York State, at the Grand Canyon, on Lake Powell, and at other locations in northern Arizona. Schaefer has suggested the deployment of a global network of such foils for global monitoring of air pollution and has stated that a global distribution of a thousand such slides could be made almost immediately.

The deposition from moving aerosols, which, of course, is the type of deposition occurring on Schaefer's foils, has been discussed in detail by Davies.[51] Large particles move across the direction of motion of the aerosol largely by sedimentation, but submicron diameter particles may deposit as a result of interception, diffusion, electrical forces, or impaction. Close to a surface, thermophoresis or diffusiophoresis may become important. And of course, there is a large difference in deposition behavior between laminar flow and turbulent flow, just as there is between still air and stirred air settling in a closed container.

Perhaps the most common of the devices for collecting particles by sedimentation from the open atmosphere are the so-called sootfall or dustfall jars. A number of varieties of these jars have been described by Jacobs.[52] A method which has been used by the Air Pollution Control District of Los Angeles County simply involves placing jars containing water (with mouth diameters of $4 \frac{5}{16}$ in.) in selected positions for 1 month. The jars are then collected, replaced, and the contents analyzed. A slightly more elaborate method, which Jacobs called the jar method, involves mounting a 1-gallon jar of glass or plastic, having a mouth diameter of 4.4 in., on a 4-ft-high wooden stand. Water or water containing antifreeze is placed in the jar which is allowed to remain on the stand, open to the air, for 1 month. The jar is then replaced and the contents analyzed. The stand and jar should be placed on a roof about four stories above the ground.

A method used in Great Britain is even more elaborate. A large glass collecting bowl is mounted above a small-mouth bottle, and a hole in the bottom of the former is connected to the mouth of the bottle. The system is used dry except for rainwater. A mesh bird shield is placed over the collecting bowl. The system has been criticized because particles deposited in the bowl may not be wetted, and thus may be re-entrained by the wind and removed from the bowl.

Several directional dustfall collection devices have been developed. Jacobs[52] derived one which was a short cylinder consisting of eight stainless steel segments forming a pielike arrangement. Mounted above the cylinder was a circular aluminum cover that had one opening of the same size and shape as one of the pie segments. The cover was mounted on a bearing so that it would turn in the wind. Thus which segment was open to receive the settling dust depended on the wind direction.

An unusual collector for particles settling from the open air has been described by Lane and Clark.[53] It was developed to collect such particles "quantitatively" for measurement of their chemical, radiochemical, and physical properties in order to characterize radioactive fallout. As the authors point out, however, watch glasses, sticky paper, grease-coated plates, metal pans, pans with Hexel inserts, and pans filled with plastic beads have all been used for collecting settling dust. Collectors are needed that prevent particle bounce-out on initial impact or blowout before particle recovery. The collector is simply a 24 × 24 in. aluminum pan with an insert made from venetian-blind louvers set at an angle of 45° from the vertical. During use the concave surface of the louvers face upward and open in the direction from which the particles arrive. At the end of each collection the louver insert is removed from the pan and either tapped or broken clear. The collector is equipped with a gasketed, clamped bed to ensure that the collector remains clean prior to exposure and that no particles are lost while returning the collector to the laboratory. When the device was used in a region of heavy rainfall (Costa Rica), it was equipped with a drain and filter to prevent particle loss in water overflowing the tray. Particles filtered from the water were combined with those remaining in the tray.

This sampler has been criticized by Lucas[54] on the grounds that although it will probably completely retain falling dust, it will only provide representative results if it is placed on the earth's surface in a region in which that surface is completely retentive. In other words, it will collect re-entrained dust on a windy day. However, according to Lucas, this effect can be minimized by mounting the collector a few feet above the earth's surface, since most of the re-entrained dust travels close to the ground.

Dustfall jars for the most part collect only those particles that are at least a few microns in diameter. Only relative significance should be attributed to the resulting data, and only then if conditions are carefully standardized. Collections of dust in such jars have been made in a number of United States cities for many years, and the results seem to indicate a trend downward in dust concentrations, perhaps because of less soft-coal burning and better control of industrial emissions. However, the incidence and intensity of smog over cities may actually be worsening since photochemical smog, which has been replacing that due to coal burning during several decades, contains largely very fine particles that are much smaller than most fly ash and flakes of soot.

Rainfall collectors can be considered to be samplers operating on the principle of sedimentation. There are many types, many of them automated, and the subject is largely beyond the scope of this book. However, one type has been used in a national precipitation network in the United States for monitoring the trace constituents, including particulate constituents, in rainfall. The results of such monitoring are of interest both from the standpoint of the composition of the rain or snow, which has implications with regard to agriculture and the biosphere in general, and because the precipitation is a sampler, although a rather poor one, of atmospheric trace constituents. Furthermore, precipitation is a major mechanism for the removal of trace constituents from the air, and estimates of the rates of such removal on a global scale are an important part of establishing the "budget" or "cycle" for some trace constituents such as sulfur compounds.[55] The collector employed is of interest in the present context if for no other reason than that it had to be designed to remain open only during periods of precipitation, thereby excluding dry fallout.[56] This was accomplished automatically with a moisture-sensing grid that activates the lid. This grid is thermostated so that snow and sleet as well as rain are collected. Furthermore, the grid is maintained at a temperature sufficient to evaporate moisture from the grid after precipitation ceases, causing the lid to close. The precipitation, which is collected in a polyethylene bucket, is recovered after each precipitation episode and mailed in polyethylene bottles to a central laboratory for analysis. The bottles contain a small amount of toluene to minimize biological activity in the sample.

The Automatic Precipitation Collector, an instrument with characteristics similar to those just described, is available from the Microchemical Specialties Co. It collects samples only during periods of actual precipitation by means of a solid-state sensing unit which can be adjusted to distinguish between heavy dew or fog and actual precipitation. Covering and uncovering the collection vessel requires no more than 15 sec before or after a period of precipitation. It is designed for long periods of unattended operation. An interesting feature is that it can be used in the reverse mode, that is, to collect dry settling particles and exclude precipitation.

Numerous variations of closed sedimentation chambers have been used, especially for laboratory studies of aerosols. Cylinders that are open at both ends can be moved back and forth through the aerosol, closed at one end (the top) and placed with the bottom on a collecting surface. Alternatively, the aerosol can be blown or drawn through the cylinder. Making the walls thick and of metal maintains a uniform temperature within the cylinder and thus decreases convection.

Green and Lane[47] point out that when sampling highly concentrated aerosols containing very fine particles, such as many smokes, the sedimentation chamber (cell) should be very shallow, for example, a few millimeters in depth, to avoid overlap on the collecting surface and coagulation during settling. The use of a shallow cell also helps to avoid errors due to convection and to avoid the need for inconveniently long settling times in order to allow all of the particles to settle out of the suspension.

The characterization of droplets collected on surfaces by any of a number of techniques, including sedimentation, presents difficulties because of spreading of the droplets and, in the case of volatile liquids, their evaporation. Specially prepared slides often help avoid such difficulties. Hydrophobic coatings, such as Vaseline, help prevent spreading of aqueous droplets and can be used to help preserve them. Dautrebande[57] suggested the following method for coating slides with Vaseline and preserving the collected droplets.

1. Use only scrupulously clean, scratch-free slides. If necessary, wash with chromate cleaning solution and soap and water, finally rinsing with alcohol.

2. Wipe the slides with a soft cloth, microscope slide tissue, or chamois skin.

3. Coat the slides with a thin layer of Vaseline, using a soft cloth.

4. Wipe off excess Vaseline with a soft cloth. Use a single smooth stroke from one end of the slide to the other. An almost invisible layer of Vaseline should remain.

5. Press a second clean, scratch-free slide onto the Vaseline-covered surface of the first and rub the slides back and forth with respect to each other. Gently warm the backs of the two slides by passing them, still pressed together, over a Bunsen flame, and then press the warm slides even more closely together using a cloth. Now pull one slide over the surface of the other along the long axis of the slides until they are separated. An even and nearly invisible film of Vaseline should remain on one surface of each slide. The slides should still be slightly warm at the completion of this operation.

6. The slides can now be placed in the sedimentation chamber. When collection of the droplets is complete, a large microscope cover slip is placed on the slide, but without pressing it into place, as this would distort the droplets. The edges of the cover slip are sealed to the slide with household cement, which can be done directly by moving the nozzle of the tube around the edge of the slip. The edges must be completely sealed, but an excess of cement should be avoided. Specimens prepared in this manner are said to last for years.

Slides can also be coated with Vaseline by painting them with a solution of Vaseline in petroleum ether and allowing the solvent to evaporate. This method is simpler than the technique described above, but it does not produce as uniform a coating. Variations of this procedure have been described in the literature.[58,59,60]

Records of the diameters of large droplets (> 5 μm radius) can be obtained by coating the collecting slides with a very soft material which the settling drops penetrate with little or no distortion. Magnesium oxide is often used for this purpose, the slides being coated by holding them over a burning ribbon of magnesium.[59] Soot is also sometimes used; the slides are coated by holding them over a benzene flame.

Farlow[61] refers to Schaefer[62] as having investigated the use for water droplets of thin films of dye bound with gelatin, polyvinyl alcohol, ethylhydroxycellulose, and other binders. Apparently this approach was not very useful because of humidity effects and poor small-droplet response.

Farlow[61] developed a durable transparent 35mm film sensitive to droplets of distilled water and seawater in the size range 1–100 μm in diameter. The film consisted of 35mm cellulose acetate film leader stock coated with polyvinyl alcohol containing silver nitrate and hydrogen peroxide. The film had the following properties:

1. The water droplets produce permanent craterlike impressions on the films by distorting the water-soluble coating.

2. The hydrogen peroxide and silver nitrate were incorporated into the coating to promote a precipitation reaction between the chloride ions of seawater and the silver ions. The silver chloride can be photochemically reduced to colloidal silver, resulting in the production of deep red-brown craters, whereas the craters produced by distilled water remain transparent.

3. The distilled water droplet craters were readily measurable down to 10 μm in diameter, but below this the resolution was difficult. However, Farlow found that the distilled water droplet craters became well-defined and colored even in the 1–10 μm range when exposed to the vapor of water, phenylhydrazine, and ammonia, although no explanation for this effect was known.

4. The films were found to be insensitive to high-activity radiation and could be used for radioautographing by coating the chemically developed, droplet-impinged film with a transparent water-repellent varnish. Photographic nuclear emulsions could then be applied directly to the film surface.

Farlow and French[63] calibrated these films with a method of more than passing interest since it can be used for calibrating any such collector, including the carbon and the magnesium oxide coated slides described above. The method is based on the assumption that the concentration of particles in a well-stirred bulk suspension and the average concentrations of droplets prepared from that suspension will be identical. Thus a count of the number of particles in an impression on the film formed by a sedimenting droplet can be used to calculate the size of the droplet before it hit the surface. Since there is a statistical variation in the number of solid particles a drop of given size will contain, many spots of a given size must be measured, and the solid particles within them counted and averaged. Farlow and French first used corn smut spores of 7.8-μm mean diameter and later polystyrene spheres of 0.514-μm mean diameter as the solid particles. They found that the film craters were about twice the size of the droplets that produced them. Of course, if such a bulk suspension can be prepared from which the droplets of immediate interest can be dispersed, collection on a plain microscope slide is acceptable.

May[64] used pure gelatin films on microscope slides for collecting and measuring fog droplets. Although the craters formed by the droplets were virtually invisible with ordinary illumination, those formed by droplets as small as 1 μm in diameter could be seen and measured using positive phase-contrast illumination. The microscpoe slides were first cleaned with hot chromic acid cleaning solution and stored in a detergent solution. A slide was prepared by wiping it dry with a clean cloth and spreading a thin film of a warm 5% gelatin solution over the slide with a glass rod. As would be expected, May found that the method was applicable to the detection of volatile droplets of other liquids, provided a clear film that was soluble in that liquid could be found. Kerosene droplets were detected with a coating of a silicone resin applied in the same manner as were the gelatin films.

Sedimenting droplets can also be preserved by allowing them to settle into a liquid of low viscosity in which they are insoluble. A difficulty is that the surface tension of the collecting liquid may prevent the droplets from penetrating into the liquid. Thus the droplets must be relatively large, have a high density, or both. It may be necessary to try a number of collecting liquids before one is found that is satisfactory. An interesting modification of this method has been applied to sprays. The collecting liquid is cooled to far below the freezing point of the droplets so that they solidify before they can coalesce. If the particles are removed to a "cold room" before they melt, they can be characterized by conventional methods such as sieving.

As mentioned above, particles much less than 1 μm in diameter are not collected at all efficiently by sedimentation techniques. A method for surmounting this difficulty is to increase the size of the particles by condensing water vapor on them. This is generally accomplished using a cloud chamber technique that involves adiabatically expanding moist air to produce supersaturation with respect to water vapor. The particles serve as condensation nuclei on which droplets form which settle to the bottom of the expansion chamber. The droplets can then be counted or allowed to evaporate leaving the original particles to 'be examined. This method was developed by Aitken in the late 1800s (hence the term Aitken particles or Aitken nuclei for particles having radii smaller than 0.1 μm), and various modifications have been used since then. The subject has been reviewed by Pollak[65] and by Mason.[66] The instrument most commonly used is a modification of a portable instrument designed by Aitken. It has a shallow test chamber the walls of which are lined with wet blotting paper that maintains the humidity in the chamber near saturation. The chamber is sealed with a plain glass top and a ruled glass bottom on which the droplets are collected. A mirror is used to illuminate the glass collecting surface. Dark-field transmitted illumination is achieved by providing the mirror with a black center, thereby providing convergent illumination. The droplets are observed with a lens mounted above the upper plate. A hand pump introduces the samples of air and produces the expansion that causes supersaturation of the air in the test chamber. The pump samples a small, known volume of the dusty air which is diluted with filtered air. A set of valves controls the pump, causing expansion, drawing in filtered air, or drawing in the sample. Instead of depending on the purity of the filtered air, particles in the air of the test chamber can be removed prior to drawing in the sample by making several expansions until no more droplets appear. If, after several expansions, droplets continue to be formed, leakage is indicated. In fact, one of the problems encountered with this device is a marked tendency to leak.

The technique is seldom used today for actual collection of the particles, but it is routinely used in conjunction with various methods, such as those based on light attenuation, to determine the concentrations of the droplets without allowing them to settle. Such techniques are described in a later chapter.

4. CENTRIFUGAL METHODS

As was mentioned in the preceding section, unless the concentrations of particles are very large, as in the case of certain smokes, sedimentation methods that depend upon gravitational forces are limited to determining the size distributions of particles about 2 μm or larger in diameter. This lower limit of particle size can be greatly decreased by using centrifugal instead of gravitational acceleration to produce the sedimentation. Since centrifugal acceleration may be many times gravitational acceleration, this change greatly increases the rate of sedimentation. Svedberg and his co-workers developed an "ultracentrifuge" that produced accelerations 750,000 times that of gravity. Such devices were used for studying the sizes of large molecules suspended in a liquid. Ordinary laboratory centrifuges can be used to determine the sizes of particles as small as 0.05 μm in diameter.

There are several ways to produce the centrifugal forces. One is to use some conventional form of centrifuge in which, under ideal conditions, the gas in which the particles are dispersed remains stationary with respect to the walls of the containers. The walls of other devices, such as cyclone separators or certain coiled tubular collectors, remain stationary, and only the aerosol moves. Some devices, designed to decrease the Reynolds number and thus the turbulence of the moving suspensions, combine these approaches.

The use of a conventional ultracentrifuge introduces a number of complications, and some quite elaborate techniques have been developed to overcome them. The acceleration varies not only with the angular velocity but also with the linear velocity and therefore with the distance of the particle from the center of rotation. In a conventional centrifuge this distance changes appreciably as the particles settle. If the centrifuge is to be used for particle-size determinations it must be smooth running to avoid convection currents, but this requirement is much less important if the device is to be used for collecting all the particles from a suspension without classification. For these and other reasons, conventional style centrifuges are seldom used for investigating the particles in aerosols, but the principles on which such centrifuges are based are often used.

4.1. Theory

First consider the case where the acceleration is produced by flowing the aerosol through a curved duct. This can be considered to be a type of inertial impaction (on the outer walls of the tube) since the inertia of the particles tends to keep them moving with a larger radius of curvature than that of the suspending gas (and the duct). The following discussion of the theory for this case is based on the development by Ranz,[8,67,68] who applied terminal velocity relationships to calculating the impaction efficiencies on the outside wall. The gas then flows with a circular motion with a constant radius, R, and a characteristic velocity, v_0. Impaction is considered to occur only from the gas close to the outside wall. If the diameter of the duct is small relative to the radius of curvature, the acceleration, a, toward the outer wall can be considered to be constant and equal to v_0^2/R. If the gas undergoes n revolutions, the particles will be subjected to a for a time $t = 2\pi Rn/v_0$. At the end of this time the depth of gas near the wall that has been cleared of particles of size d will be $2\pi vRn/v_0$, where v is the terminal velocity of the particles when subjected to the acceleration, a. The Ranz development assumes that the particles attain terminal velocity at the start of the process.

Ranz now defines the impaction efficiency, E_I, as the depth of the gas layer cleared of particles of size d divided by the total thickness of the layer, that is, the width of the duct:

$$E_I = \frac{v2\pi Rn}{v_0 D_c} \qquad (2\text{-}61)$$

where D_c is the width of the duct.

Now, when the Stokes equation applies [equation (2-51), with a substituted for g],

$$E_{IS} = \frac{2\pi n\rho v_0 d^2}{18\eta D_c} \qquad (2\text{-}62)$$

For the more general case, $24/\psi Re$ is calculated as a function of $24\sqrt{2}\ N_D{}^3\ (D_c/R)N_{\rho g}$ where ψ is the drag coefficient, here defined as (drag force)$/(\rho' v^2/2)A$; A is the particle area projected in the direction of motion; N_D is the square root of the inertial parameter, equation (2-8), and is defined as $(\rho' v_0/18\eta D_c)^{1/2}d$; and $N_{\rho g}$ is $9\rho'^2 v_0 D_c/\rho\eta$. The results of such a calculation are shown in Figure 2-15, which can be used to estimate the maximum possible collection efficiencies of particle collectors such as cyclone separators, centrifuges, and screw separators.[69]

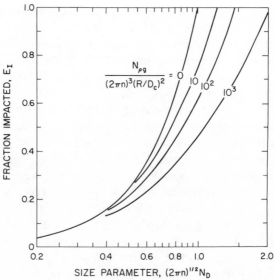

Figure 2-15.   Impaction on the outside walls of a sweeping bend or vortex.[67]

The figure shows the fraction of particles of size d which impact on the outer surface of the duct. Curves are shown for various values of the parameter $N_{pg}/(2\pi n)^3 (R/D_c)^2$, which accounts for air inertia (represented by ρ') and does not depend on the particle size. When the Stokes equation applies, the value of this parameter is effectively zero. Note that all particles above a certain size hit the collecting surface, but that the collection efficiency decreases rapidly with decreasing size for smaller particles. Ranz emphasizes that Figure 2-15 is for a highly idealized system and does not apply when D_c approaches R in size, since in this case v_0 is far from constant and extensive numerical calculations of limiting particle trajectories are required.

Davies[19] has discussed the deposition in a cyclone collector, which is shown schematically in Figure 2-16. It is a device with no moving parts that consists essentially of a chamber in which the gas stream velocity is converted to a swirling motion. The resulting centrifugal acceleration moves the airborne particles to the wall.

There are several types of cyclone collectors, but perhaps the most common is that shown in Figure 2-16. It has a tangential inlet near the

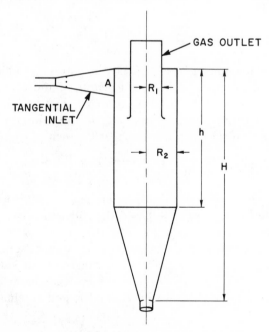

Figure 2-16. Schematic diagram of common cyclone collector.

top of a vertical cylindrical chamber. The gas outlet is a cylinder about one-half the diameter of the chamber, mounted concentrically in the top of the chamber, and extends about one-third of the way down. The particles approaching the wall of the chamber are led into a cone-shaped extension of the chamber, which in turn leads the collected particles down to an oval discharge port. The entering aerosol spirals downward to below the gas outlet, and near the bottom of the cone reverses direction to form an inner vortex which is generally smaller in diameter than the gas outlet. A layer of dust-rich gas, produced by the centrifugal force, spirals down to the discharge port.

Because of their simplicity and ease of operation, cyclones of various forms are used at times for sampling aerosol particles, but they are more useful for industrial applications. They are nearly useless for collecting submicron-diameter particles but have the great advantage over static sedimentation chambers of being able to sample a large amount of air in a short time.

A relatively simple treatment of cyclone theory for an idealized gas has been described by Green and Lane.[47] The following is an equation developed by Davies[19] for the time, Δt, for a particle to travel from a radial distance R_1 to R_2 in a gas density ρ' and viscosity η (see Figure 2-16):

$$\Delta t = \frac{(R_2{}^4 - R_1{}^4)\, 9\eta}{8r^2(\rho - \rho')\, v_0{}^2 R_2{}^2} \qquad (2\text{-}63)$$

where v_0 is now the tangential air velocity at radius R_2. This is the time required for the particle farthest removed from the chamber wall to reach that wall. Now, if H/v_0 is substituted for dt, which for obvious reasons is not entirely justified, and $r_{\min}$ is the minimum size above which the collection efficiency would be 100%, then

$$r_{\min} = \left[\frac{(R_2{}^4 - R_1{}^4)\, 9\eta}{8R_2{}^2 H v_0(\rho - \rho')} \right]^{1/2} \qquad (2\text{-}64)$$

Green and Lane state that this equation has been criticized on the grounds that as $R_1 \to R_2$ $r_{min} \to 0$, and the collection efficiency becomes infinite. However, Davies has responded that as $R_1 \to R_2$ the volume flow rate must approach zero, and no particles would escape.

Numerous other equations have been suggested for estimating the relationship between particle size and the cyclone characteristics, such as the following proposed by Rosin, Rammler, and Intelmann:[70]

$$d_{cp} = \left(\frac{9\eta w_i}{2\pi n v_i(\rho - \rho')} \right)^{1/2} \qquad (2\text{-}65)$$

where d_{cp} is the diameter of the particles collected with 50% efficiency, w_i is the inlet width, n is the effective number of turns in the cyclone, and v_i is the gas inlet velocity. Cyclones are most useful for particles in the size range 5–200 μm diameter. Smaller particles are usually not collected efficiently, and larger ones are better collected by sedimentation.

Equation (2-65) has been criticized on the grounds that it assumes a constant centrifugal force to establish the velocity pattern, whereas, as was described above, in most cyclone collectors the gas path is a double spiral. There may also be an eddy at the top of the cyclone, and this may contribute to the removal of particles.

The venturi scrubber is another device that is used primarily for the cleaning of industrial effluents, but is used occasionally for collecting samples of particles suspended in gases. It combines the scrubbing action of water droplets moving through an aerosol with the collection of the droplets using a cyclone collector. It has the advantage that a large volume of air can be sampled with a small volume of water, since the water is recycled. Figure 2-17 is a schematic drawing of a portable venturi scrubber used in some of the early studies of smog in Los Angeles and described by Magill et al.[71] Air is drawn through the device at the rate

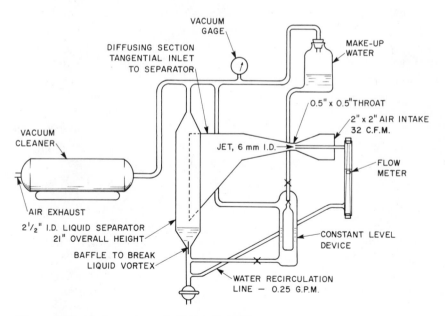

Figure 2-17.   Laboratory-scale Venturi scrubber.

of 32 cfm by a tank-type vacuum cleaner. A high-volume sampler pump and motor will do very well. The scrubbing liquid is sprayed into the venturi throat as a result of the large pressure drop occurring there. At the venturi throat the gases have a velocity of 250–300 ft sec^{-1}. The scrubbing action of the droplets occurs from the time they are produced until they have been removed by the cyclone separator.

The scrubbing liquid is recirculated from the bottom of the separator back through the jet in the venturi throat. A satisfactory water jet tube was found to be a length of glass tubing, 6 mm in inside diameter, with a straight-cut end. This end is beveled toward the center to produce a sharp edge. A reservoir and constant-level device are used to make up evaporation losses. An area-type flowmeter in the water-circulation line was used to assure the maximum possible flow rate, which gave optimum scrubbing efficiency. This flow rate was equivalent to 8–9 gal of water per 1000 ft^3 of air treated. The vacuum gage shown in the figure can be calibrated to show the airflow rate. The width of the diffusing section where it is attached to the cyclone separator is about 0.75 in.

The velocity of air through the venturi throat and the liquid-to-air ratio were comparable to those of industrial scrubbers. The collection efficiency of the device was determined for several types of particles and gases, using several scrubbing liquids.

The theory of collection by venturi scrubbers is quite complicated. The gas in the venturi throat breaks the liquid into droplets having a wide size distribution, and since there is initially a large velocity differential between these droplets and the gas with its suspended particles, collection by impaction occurs. The droplets are accelerated to the gas velocity, but in the cyclone collector there are again differential velocities dependent on particle and droplet size that can result in impaction of any remaining particles by the droplets. Impaction is, indeed, the principal collection mechanism, and the following empirical equation was developed by Johnstone, Feild, and Tassler:[72]

$$E = 1 - \exp\left(-kLK^{1/2}\right) \tag{2-66}$$

where E is the collection efficiency, k is a constant, L is the liquid-to-gas ratio, and K, as usual, is the inertial parameter at the venturi throat.

The venturi scrubber, although extensively used in industry, is seldom used for atmospheric sampling, which is perhaps unfortunate because of its versatility (it can be used for both particulate matter and gases) and its high sampling rate. It is readily constructed in a shop; the one shown in Figure 2-17 was made of Lucite.

A basic equation that is applicable to conventional centrifuges of the types used for suspension in liquids is a modification of Stokes' equation that was developed by Svedberg:

$$d = \frac{6}{\omega} \left[\frac{\eta \ln (s_2/s_1)}{2(\rho - \rho') t} \right]^{1/2} \qquad (2\text{-}67)$$

where s_2 is the distance from the axis of rotation to the bottom of the centrifuge tube, s is the distance from the axis of rotation to the part of the suspension closest to the axis of rotation, t is time, and ω is the angular velocity. For any given set of experimental conditions this equation reduces to the form

$$d = \frac{k}{\sqrt{t}} \qquad (2\text{-}68)$$

Centrifuges in this form are not used for collecting particles from gases, but these equations may be applicable to some types of gas centrifuges.

4.2. Instrumentation

Almost all centrifuges for aerosols that include spinning parts are designed to provide classification of the particles according to size as well as removal of the particles from the gas phase. These are described in a later chapter. However, there are a few exceptions, one of which is the Conicycle designed by Wolff and Roach[73] to collect particles in a specific size range. It is battery operated and consists essentially of a motor-driven sampling head which rotates at 8000 rpm. Its deposition behavior is intended to be similar to the British standard for deposition in the human respiratory system; thus it prevents particles above a predetermined diameter from entering the sampler, collects all the particles in a predetermined size range, and passes those below the lower limit of complete collection to an extent inversely proportional to their diameter.

In addition to the small venturi scrubber including a cyclone, which has already been described, various miniature cyclone collectors have been used for sampling large particles from suspensions. Figure 2-18 shows the dimensions of an all-glass cyclone collector described by Kanter, Lunche, and Fudurich[74] and used to collect large particles among the air contaminants from combustion sources. All the inlets and outlets, including that at the bottom of the cyclone which was attached to a particle receiver, were equipped with ground-glass ball joints. It

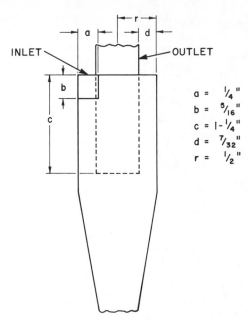

Figure 2-18.   Miniature glass cyclone. Reprinted with permission from *J. Air Poll. Control Assoc.* **6,** 191 (1957).

was usually backed up by an Alundum thimble filter and finally an impactor, the filter collecting particles passed by the cyclone and the impactor ("impinger") collecting particles passed by the filter.

Numerous spiral and helical devices have also been designed. They are usually made of glass tubing and sometimes coated on the inside with an oily substance such as silicone oil to increase retention. At the completion of a sampling period the particles can be removed by flushing with a solvent for the oil. Like the cyclone, such instruments are only effective for large particles and are often used in conjunction with some device for collecting smaller particles. The collection efficiency of such a device can be roughly estimated from the theory presented above.

Stevens and Stephenson[75] have developed a commercially available single-stage selective particle sampler which has a particle size-deposition curve similar to that of the pulmonary compartment of the lung as proposed by the International Commission on Radiological Protection Task Group on Lung Dynamics.[76] The report of the task group presents the theoretical deposition of unit density spheres in three parts of the respiratory system, namely, the nasopharyngeal, tracheobronchial, and pulmo-

nary, at tidal volumes of 750, 1450, and 2150 cm³ and at a breathing rate
of 15 respirations per minute. This theoretical deposition did not differ
seriously from the available experimental data on lung deposition, so
the theoretical data can be used for hazard assessment. The report also
suggests that a tidal volume of 1450 cm³ at 15 respirations per minute
may be considered to be the mean. On this basis, the problem of assessing
the hazard from airborne particles is reduced to the determination of
the amount of such particles that would be deposited in the nasopharyngeal and pulmonary compartments under a single respiratory pattern.

The deposition curves are shown in Figure 2-19 and a schematic
drawing of the sampler in Figure 2-20. The aerosol is drawn through an
annular slot; at point x part of it is turned through a sharp bend, drawn
through another slot, and then through the central portion of a glass
fiber filter paper. At the sharp bend, particles with sufficient inertia to
cross the airstream enter a second region of air flow and are carried to
the outer annular section of the filter paper. According to the authors,
any type of filter paper can be used, provided it will allow a flow rate of
2 liters per minute and that a good pressure seal can be produced at the
filter paper between the central and annular regions. A smooth air flow
is required, and a diaphragm pump equipped with a flow pulsation
damper such as that described by Harris and Maguire[77] is recommended.
The aerodynamic diameter of Figure 2-19 is the diameter of a spherical
particle of unit density that settles at the same rate as the particle being
considered.

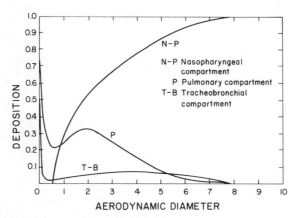

Figure 2-19. Lung deposition as a function of particle size for 15 respirations
min⁻¹, 1450 cm³ tidal volume. Reprinted with permission from Stevens, D. C.,
and J. Stevenson, *Aerosol Sci.* **3,** 15 (1972), Pergamon Press.

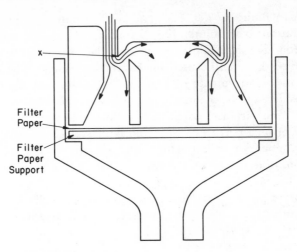

Figure 2-20.   Design of a size-selective sampler. Reprinted with permission from Stevens, D. G., and J. Stevenson, *Aerosol Sci.* **3,** 15 (1972), Pergamon Press.

The sampler was calibrated with monodispersed aerosols of polystyrene spheres that were radioactively labeled with Cr-51, using chromium acetyl acetonate. Activity was measured on the central and annular regions of the filter paper and on the internal surfaces of the sampler. Judging from the aerodynamic diameter-activity curves, the amount of dust deposited in the center of the filter can be used to estimate the concentrations of dust that will be deposited in the pulmonary compartment of the respiratory system. Furthermore, the amount of particulate material on the outer, annular region of the filter can be used as a measure of nasopharyngeal deposition. The sampler design can be modified to conform to future changes in respiratory deposition model curves.

Stevens and Churchill[78] have described another version of the above sampler that has a much higher volume flow rate (35 liters per minute). It is also commercially available. The fiber filter paper for this model is 6 cm in diameter. It has essentially the same deposition characteristics as the earlier model.

Unfortunately, the pipes and tubing used to conduct aerosols from probes to sampling equipment often have sharp bends that make them quite efficient centrifugal collectors. The theory described above can be used to estimate errors from this source. The design of sampling probes and leads is discussed in detail in a later chapter.

5. IMPACTION

A widely used technique for collecting airborne particles prior to examination is known as impaction or impingement, or sometimes inertial impaction. The principle is the same as the filtering mechanism of impaction, described earlier; namely, when a moving aerosol changes direction, the particles tend, because of their inertia, to continue in the same direction they were going. If the change in direction of the aerosol was caused by a diverting body (a fiber in the case of fiber filters), some of the particles may impinge upon that body and be collected. A particle-collecting device based on this principle can be designed to collect a single deposit of particles with each operation or to classify the particles into several fractions according to size.

The impaction can be produced in several ways. For instance, the object on which the particles are to be collected may be moved rapidly through the aerosol by mounting it on the outside of an airplane or whirling it with a motor. The object may be a sphere, fiber, plate, rod, or a collection of these. A fiber or other object can be suspended in wind, natural or artificial, and the wind will provide the differential velocity. One of the most common arrangements is to force the aerosol to flow at very high velocity, often sonic, through an orifice (which may be circular or a slit) and allow the emerging air to impinge on a flat surface. The air stream is sharply diverted and some or all of the particles are collected.

5.1. Theory

The theory of impaction on cylinders, which includes the theory of impaction on cylindrical fibers, has been discussed earlier in connection with filtration. Ranz and Wong[8] have given the following equation for impaction from an aerosol stream of infinite extent on a ribbon of infinite length and width, D_c:

$$E = \frac{S_2 - S_1}{S_2 \exp (S_1 f') - S_1 \exp (S_2 f')} \tag{2-69}$$

where

$$S_{1,2} = \frac{1}{4K} \pm \left[\left(\frac{1}{4K} \right)^2 + \frac{1}{2K} \right]^{1/2}$$

$$f' = \frac{1}{q} \tan^{-1} \left[\frac{4Kq}{4K - 1} \right]$$

$$q = \left[\frac{1}{2K} - \left(\frac{1}{4K} \right)^2 \right]^{1/2}$$

Figure 2-21 is a plot of this equation and also one for impaction on a disk.

Ranz and Wong gave the following equation for impaction from a rectangular aerosol jet of infinite length and width D_c on a flat plate of infinite extent:

$$E = \frac{S_2 - S_1}{S_2 \exp (S_1 f') - S_1 \exp (S_2 f')} \tag{2-70}$$

where

$$S_{1,2} = -\frac{1}{4K} \pm \left[\left(\frac{1}{4K} \right)^2 + \frac{1}{2K} \right]^{1/2}$$

$$f' = \frac{1}{q} \tan^{-1} \left[\frac{(1 - 2K)\, 8Kq}{4(K - 1) + (4Kq)^2} \right]$$

$$q = \left[\frac{1}{2K} - \left(\frac{1}{4K} \right)^2 \right]^{1/2}$$

For impaction for a round aerosol jet on a flat plate of infinite extent they gave the equation:

$$E = \left[\frac{S_2 - S_1}{S_2 \exp (S_1 f') - S_1 \exp (S_2 f')} \right]^2 \tag{2-71}$$

where

$$S_{1,2} = -\frac{1}{4K} \pm \left[\left(\frac{1}{4K} \right)^2 + \frac{1}{2K} \right]^{1/2}$$

$$f' = \frac{1}{q} \tan^{-1} \left[\frac{(1 - 4K)\, 8Kq)}{(8K - 1) + (4Kq)^2} \right]$$

$$q = \left[\frac{1}{K} - \left(\frac{1}{4K} \right)^2 \right]^{1/2}$$

Plots of equations (2-70) and (2-71) are shown in Figure 2-22.

Ranz and Wong also produced experimental efficiency curves for inertial impaction from both rectangular (slit) and round aerosol jets. The method was similar to their study of the efficiency of particle collection by impaction on fibers described in the section on filtration. The aerosol was generated in a condensation generator with glycerol as the aerosol material. The particle diameters ranged from 0.3 to 1.4 μm.

The rectangular jets had cylindrical, convex inside walls, and the round jets had a 60° cone for the approach line. The rectangular jets were from 10 to 50 times longer than wide. The distances between the jets and the collector plate were from one to three jet widths or diameters away. The collector plates were actually the bottoms of flat-

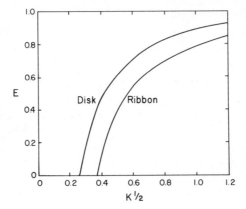

Figure 2-21. Theoretical impaction efficiencies of disk and ribbon. Reprinted with permission from Ranz, W. E., and J. B. Wong, *Ind. Eng. Chem.* 44, 1371 (1952). Copyright by the American Chemical Society.

bottomed glass cups which were weighed before and after impaction. Particles that escaped impaction were collected by filtration and weighed.

The measured efficiencies were somewhat smaller for a given value of K than predicted by theory and the experimental E versus $K^{1/2}$ curves had more of an S shape. As predicted by theory, the round jet was somewhat more efficient than the rectangular one. The experimental values

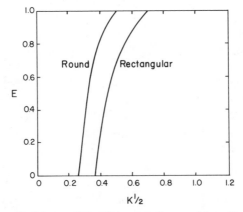

Figure 22. Theoretical impaction efficiences of aerosol jets. Reprinted with permission from Ranz, W. E., and J. B. Wong, *Ind. Eng. Chem.* 44, 1371 (1952). Copyright by the American Chemical Society.

for $K^{1/2}$ for 50% efficiency were 0.57 for rectangular jets and 0.38 for round ones. Because the ratio of particle diameter to jet diameter was very small, collection by interception could be considered to be negligible. The aerosol particles were found to be uncharged; thus electrostatic collection was also improbable.

Marple and Liu[79] have reexamined the theory of rectangular and round impactors and have experimentally verified their theory. They found that the jet throat length has a negligible influence on the collection efficiency but that the Reynolds number has a substantial effect below a value of about 500. The effect of the jet to collecting plate distance is small above a value of 1.5 jet widths and 1.0 jet diameters for rectangular and round impactors, respectively. Marple and Liu recommended that impactors be designed with jet to plate distances exceeding these critical values.

When particles are small relative to the mean free path of the suspending gas, K must include the Cunningham correction factor in the numerator. Thus the collection efficiencies of impactors for small particles increase with decreasing pressure in the particle size and pressure range, where this correction becomes important.

The question repeatedly arises about the extent to which particles are retained once they come in contact with a surface. This question is especially pertinent for collection by impaction, where the particles may approach the surfaces at sonic velocities. The pragmatic approach often taken is to coat the collecting surfaces with some sticky substance, such as an oil, to retain solid particles. This method has already been mentioned several times, for example in connection with filtration. Small solid particles do attach strongly to solid surfaces through the action of intermolecular attractive forces, but they are not always captured or retained upon striking a surface. Even micron-size droplets may bounce when striking a smooth surface at high velocities.[80]

Dahneke[81] has discussed the mechanics of a solid-particle and solid-surface collision and has derived an equation relating reflected and incident velocity. There are two parameters in the equation: the coefficient of restitution, e, and the particle-surface interaction energy, E_{ps}. He considers a particle moving toward a surface but still beyond the influence of the surface with an incident normal velocity, V_{ni}, and a corresponding kinetic energy, KE_{ni}, which is regarded as the total system energy throughout the collision. As the particle nears the surface the particle falls into the particle-surface potential well of depth E_{ps}. If the rebound energy is insufficient to escape the potential well, the particle will be captured. Because the depth of the potential well may vary during the collision, for example, as a result of contact charging or of surface deformation, Dahneke defines two potential well depths: E_{psi} is the depth for the incident particle, and E_{psr} is that for the reflected particle. He then defines the coefficient of restitution, e, as the ratio of normal particle velocity at the moment of impact. Then the kinetic energy of the particle at the moment of rebound is

$$KE_{\text{at rebound}} = (KE_{ni} + E_{psi})\,e^2$$

The final kinetic energy of the reflected particle, after it has left the potential well and is beyond the influence of the surface, is

$$KE_{nr} = (KE_{ni} + E_{psi})\,e^2 - E_{psr} \qquad (2\text{-}72)$$

This can be equivalently expressed in terms of particle velocity and mass m as

$$\frac{V_{nr}}{V_{ni}} = \left(e^2 - \frac{E_{psr} - e^2 E_{psi}}{mV_{ni}^2/2}\right)^{1/2} \qquad (2\text{-}73)$$

The borderline between particle capture and escape by bouncing occurs for KE_{nr} of equation (2-72) equal to zero; thus the particle capture limit V_{ui}^* is given by

$$V_{ni}^* = \left[\frac{2}{me^2}\,(E_{psr} - e^2 E_{psi})\right]^{1/2} \qquad (2\text{-}74)$$

Capture occurs when V_{ni} is less than V_{ni}^*

Dahneke distinguishes two special cases. When the surface on which the particles impinge is thick and hard, E_{psi} and E_{psr} can be considered to be equal and to equal E_{ps}. Then

$$V_{ni}^* = \left(\frac{2E_{ps}}{m}\frac{1-e^2}{e^2}\right)^{1/2} \qquad (2\text{-}75)$$

If E_{ps} is much smaller than E_{psr},

$$V_{ni}^* = \left(\frac{2E_{psr}}{me^2}\right)^{1/2} \qquad (2\text{-}76)$$

The depth of the potential well for two contacting spheres of diameter d_1 and d_2 is

$$E_{ps} = \frac{Ad'}{12z_0} \qquad (2\text{-}77)$$

where A is the Hamaker constant, z_0 is the equilibrium separation of the spheres, and d' is the reduced diameter, $d_1 d_2/(d_1 + d_2)$. Equation (2-77) is also applicable to a sphere on a plain surface, since then $d_2 = \infty$ and d' is the diameter of the sphere. Combining equations (2-75) and (2-77),

$$V_{ni}^* = \left[\frac{A(1-e^2)}{\pi z_0 \rho e^2}\right]^{1/2}\bigg/ d \qquad (2\text{-}78)$$

With values of the Hamaker constant for quartz-quartz of 8.5×10^{-13} ergs and for polystyrene-quartz of 7.5×10^{-13} ergs, the captive limits shown in Figure 2-23 were calculated. Dahneke concludes that because particles being collected by impactors attain velocities of the order of 100 m sec^{-1}, the only way to assure the capture of solid aerosol particles by impaction is to coat the impaction surface with a viscous oil.

Jordan[82] undertook an experimental investigation of the adhesion of particles to surfaces, and he also derived equation (2-76), concluding that a solid quartz particle will stick only if $V_{ni} < (30/d)$ where d is in micrometers. This finding is a much higher limiting velocity than is predicted by Figure 2-23, but it does not change the conclusion with regard to impactors. The experimental work involved allowing particles to settle onto microscope cover slips and determining the velocity of a jet of air required to detach the particles of various sizes. Some of the results are shown in Figure 2-24.

Of course, intermolecular forces are not the only ones that can produce adhesion; electrostatic forces, for example, may contribute. However, the implications of all of this are clear. If solid particles are to be collected by impaction where velocities approach the speed of sound, coating the collection surfaces with a sticky substance is essential.

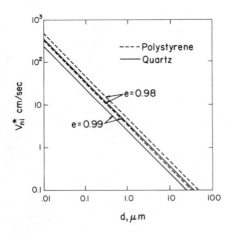

Figure 2-23. Particle capture limits for polystyrene and silica (quartz) spheres colliding with a rigid quartz surface, i.e., the surface of a thick body. Particle capture occurs when V_{ni} is less than V_{ni}^*.[81]

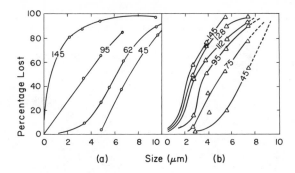

Figure 2-24.   Percentage of particles removed at various air speeds.[82] Copyright by The Institute of Physics. (a) glass dust, (b) quartz dust. Air speeds down the jet are indicated in m/sec on each curve.

One way in which particles collected by impaction may be reentrained occurs when so much particulate matter has been collected that it piles up. The collected material may then behave as a loose aggregate and slough off the collecting surface.[48]

5.2. Instrumentation

Probably the simplest device for collecting particles by impaction is simply a very fine fiber mounted on some type of holder. A very ingenious method for preparing the fibers, though for obvious reasons not a popular one today, was used by Dessens.[83] He first took a *fairly* small spider and with it made a grid of fine threads on a small metal frame. He then allowed a *very* small spider to spin a web of even finer fibers across the grid. In this way he obtained fibers that were possibly as small as 0.01 μm diameter. The frame containing these fibers was "mounted on a revolving turntable."

Occasionally spider webs are still used. For example, Rinehart[84] used such fibers for observing the change of droplets containing a dissolved salt into dry particles at various relative humidities. She designed a chamber through which humid air was passed and in which the particle suspended on a fiber could be observed with a microscope. The floor of the chamber, shown in Figure 2-25, was a microscope slide, and the spider web with its suspended particle was stretched across the slide, fastening it at each end with double-stick tape. The relative humidity was determined with wet and dry thermocouples. The use of flowing air permitted the use of an unsealed chamber if desired.

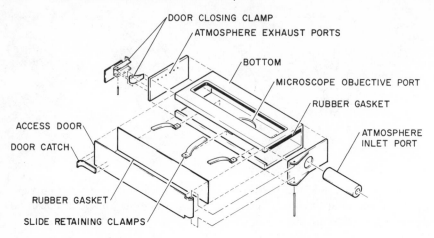

Figure 2-25.   Assembly drawing of microscope atmosphere chamber.[84]

A source of single fibers in most laboratories in which studies of aerosols are undertaken is fiber filters from which individual fibers can be drawn. The fibers in most filters, unfortunately, are not as fine as those produced by very small spiders. Another approach is to prepare fibers by dissolving a plastic material such as polystyrene in a volatile organic solvent such as methylene chloride, spreading a thin layer of the solution on a surface, and pulling out threads from the layer with another surface such as a glass rod. Some experimenting is required with concentrations

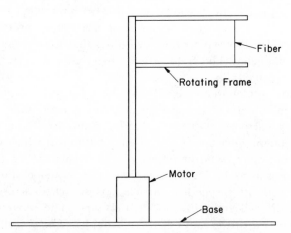

Figure 2-26.   Schematic drawing of a sampler using impaction on a single fiber.

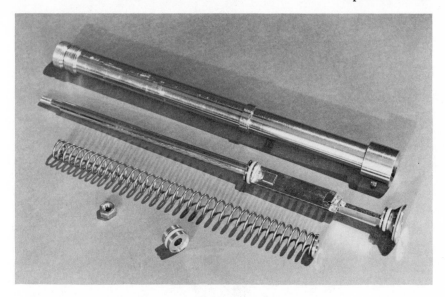

Figure 2-27. Photograph of an impactor for use on high-altitude aircraft. The impactor is disassembled and the motor assembly for pushing the impactor surfaces into the airstream is not shown. The impactor housing, shown at the top of the photograph, is 28 cm long.

of the plastic in the solvent (5–10% is usually appropriate), thickness of the solution layer, and so forth. Satisfactory fibers have been produced merely by placing a drop of solution on a forefinger, pressing thumb and forefinger together, and then pulling them apart.

Very small, battery-operated motors (2 or 3 cm in each dimension) can be used to move the fiber through the aerosol. A simple arrangement is shown in Figure 2-26. The "Rotorod" sampler is a commercial version of this device sold by Metronics Associates, Ltd., and is available as a portable table model, a model for sampling from balloons, and a model with a timed sampling sequence. It uses coated collector rods rather than fibers.

Another type of impactor which depends for its action on being impelled through the aerosol is shown in Figures 2-27 and 2-28. Several versions have been constructed, among them the one described by Junge and Manson[85] and that (shown in the figures) constructed at the National Center for Atmospheric Research (NCAR). These impactors were designed to be flown on U.S. Air Force aircraft with very-high-altitude capability to collect particles from the stratospheric aerosol layers. The

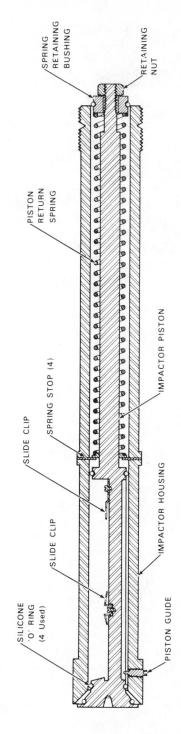

SPRING
RETAINING
BUSHING

RETAINING
NUT

PISTON
RETURN
SPRING

IMPACTOR PISTON

SLIDE CLIP

SPRING STOP (4)

SLIDE CLIP

IMPACTOR HOUSING

SILICONE
'O' RING
(4 Used)

PISTON GUIDE

Figure 2-28. Drawing of details of an impactor for use on high-flying aircraft. The impactor surface is normal to the plane of the drawing.

108

collection surfaces of the impactors must be cleaned and otherwise prepared in the laboratory (preferably a "clean room") and then sealed to prevent contamination, the seal to be broken only at the beginning of the sampling period and formed again immediately following sampling. (Variations among the impactors have been largely in the construction of the seals.) Collection was on two flat surfaces facing into the airstream (Figure 2-27). They were of different widths so as to have different impaction efficiencies. The Junge and Manson device differed from the NCAR impactor in that only the narrower tip plate extended beyond the end of the housing, the wider plate being exposed in a window in the housing. The housing for the NCAR device is shown at the top of Figure 2-27. It is actually a tube containing the rod and spring, and is mounted on the outside of the aircraft. A sample is taken by pushing the collecting surface out of the housing by a motor and into the airstream provided by the speed of the aircraft. Compression of the spring when the collection surface is extended is used to retract the surface at the conclusion of the sampling period and to provide a positive seal.

Junge and Manson used a very elaborate technique for preparing and cleaning the collection surfaces. The washed tip plate was coated with germanium or gold, by vacuum evaporation when appropriate. This plate was sometimes left dry, but to increase the retention of particles it was sometimes dipped in a 1 mg/ml solution of 250,000 centistokes silicone 200 fluid in hexane. Evaporation of the hexane left a smooth coating of the silicone. The window plate was used by Junge and Manson to collect larger particles for individual observation. It was polished with $\frac{1}{4}$-μm diamond paste to remove pits and scratches and then washed with organic solvents for the paste binder, followed by a detergent solution or chromic acid cleaning solution. Then came a hot-water rinse and a distilled-water rinse, two of the most important parts of the process, followed by draining on a lens tissue and drying in a dust-free atmosphere. The surfaces were then coated with gold or germanium and finally with silicone. The metal coating was required for analysis of the collected particles with electron microprobe techniques. The impactors were heat sealed in a polyethylene bag before being shipped.

Collections on the NCAR sampler were made on strips of heavy platinum foil mounted on the collection regions of the impactor and having the same length and width as those regions. These strips were cleaned by washing them with absolute alcohol, then with an aqueous soap solution, and finally rinsing utrasonically with distilled deionized water. At times the strips were coated with the same grade of silicone used by Junge and Manson. They were handled with forceps that had been ultrasonically cleaned with absolute alcohol. After each flight the strips were

Figure 2-29. Impactor for collecting particles on several electron microscope grids, opened to show the collecting surface.

removed from the impactors, and portions a few millimeters in length cut from each end of each strip to eliminate end effects on the collection efficiency.

The Junge and Manson impactors were flown on Lockheed U-2 aircraft, and the NCAR impactors were flown on RB-57F aircraft.

Numerous single-stage impactors have been developed in which the aerosol is drawn through an orifice at high velocity, producing a jet of aerosol that is made to impinge on a surface such as a microscope slide. The aerosol is usually drawn through the orifice with some sort of vacuum pump.

When the particle concentrations are large, as in dusty atmospheres, and the particles to be collected are relatively large, a simple hand pump can be used and all parts of the impactor incorporated in a single unit. The "conimeter" is such a device which was designed for use in mines in which a lightweight sampler was required. The unit includes a microscope for counting the collected particles. The pump is a single-stroke steel piston that draws a known volume of air (2.5 or 5 cm³) through a conically shaped jet. The collecting surface is a glass disk with a coating of adhesive to improve particle retention. The disk can be rotated beneath the jet so that up to 36 samples can be collected on a single disk.

Figure 2-30. Impactor for collecting particles on electron microscope grids, showing selection knob.

The Sartorius conimeter is constructed largely of brass, weighs about 1 lb., and its dimensions are roughly 10 × 4 in. The microscope is equipped with a focusing tube and with a net micrometer reticule for counting the particles. A dark-field condenser is also furnished for the illumination of small particles. A method for rapidly estimating the number concentration of the particles is provided by the manufacturer and consists of a series of photomicrographs of counted samples. The samples observed with the microscope are compared with the photomicrographs.

A variation of the conimeter has been used by the author and his associates that has no microscope and uses an electrically driven vacuum pump instead of the hand pump. It is used for sampling relatively uncontaminated air and has the advantages that a higher linear velocity through the jet can be achieved (and thus a greater collection efficiency for small particles), and that a very much larger volume of air can be sampled. Figures 2-29 and 2-30 are different views of this impactor. The left side of Figure 2-29 shows the inlet and exit, and the right side shows the collecting surfaces, which are indented so as to hold electron microscope grids. Figure 2-30 shows the knob for selecting the grid on which to collect the sample.

A number of impactors have been designed for collecting airborne bacteria and similar microorganisms. Several of these are described in *Air Sampling Instruments*.[33] An example is the Casella Slit Sampler in which air is drawn through a slit onto an agar culture plate while the plate is uniformly rotated, spreading the bacteria over the plate. Two basic models are available which differ in size and flow capacity, the larger unit being used to sample very clean air. The plates are rotated with a motor through a gear train, and three rotation speeds are provided. With the larger unit it is claimed that one bacterium per 100 ft^3 of air can be detected.

A device for much the same purpose is sold by the Anderson Samplers and Consulting Service. In this device the bacteria are impacted on an agar-coated rotating drum that moves downward as it revolves, producing a line deposit. The drum can be rotated at various rates, and various flow rates can be achieved by changing the jet and the pumping rate.

A number of single-stage impactors have been developed for collecting radioactive particulate matter. The impactor shown in Figure 2-31 (the Staplex annular impactor) was developed to collect particles emitting alpha particles, such as uranium, without collecting particles containing radon and thoron progeny. Its operation is based on the finding that, for the most part, plutonium is associated with relatively large, high-density particles, whereas the radon and thoron daughter (progeny) activity is usually associated with very small particles. The slit is annular and the device is somewhat like that of Stevens and Stephenson,[75] described in the section on centrifugal removal. The editor of *Air Sampling Instruments*[33] points out that small plutonium particles are also rejected and that although they may represent only a small part of the total mass of

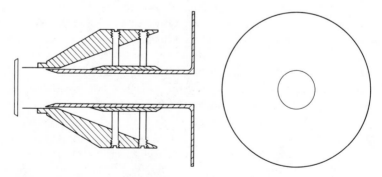

Figure 2-31.   Annular impactor.

such particles, they may be of great physiological importance. A very similar device is manufactured by C. F. Casella & Co., Ltd., of London.

Simple, continuous sampling, single-stage impactors can be constructed in most shops. For example, a jet can be mounted facing a drum driven by clockwork. The jet is built into the side of an evacuated vessel containing the drum and the clockwork. A two-quart pressure cooker makes a convenient vacuum chamber. If the particles are to be studied with a microscope, they can be collected on a strip of transparent tape wrapped about the drum. After sampling, the strip can be removed, cut into convenient lengths, and mounted on microscope slides.

High-velocity impactors have the disadvantage that the expansion of air through the jet may cause condensation of moisture if the humidity is high. A result can be the agglomeration of particles, and soluble crystals may lose their original shape and size when the water evaporates.

Impaction under liquids is a common collection technique, and the devices for doing this are commonly called impingers. One of the best known of these is the Greenburg-Smith impinger, shown schematically in Figure 2-32. It consists of a tall glass cylinder containing a vertical tube, the lower end of which narrows down to a jet 2.3 mm in diameter and 5 mm from the flat bottom of the cylinder.[86] An inch or more of

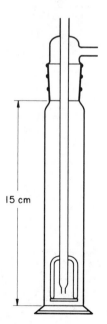

15 cm

Figure 2-32.   Greenburg-Smith impinger.

dust-free liquid, usually water, is added to the cylinder. Suction is applied to the tube at the top, with the result that the air to be sampled is drawn through the jet at the rate of 1 ft^3 min^{-1} and impinged against the bottom. Particulate matter is retained in the water. The collection efficiency can be improved by using two or more impingers in series. Suction is usually applied with a motor-driven vacuum pump, although a hand pump is sometimes used.

Schadt and Cadle[48] determined the collection efficiency of a single Greenburg-Smith impinger for particles of various size, consisting of sodium chloride, glycerol, or stearic acid. The inertial parameter values (K) were changed by varying both the particle size and the aerosol flow rate. Some of the results are shown in Figures 2-33 and 2-34.

The midget Greenburg-Smith impinger[87] is similar to the instrument just described but is about one-third the size. The flow rate is 2.8 liters of aerosol per minute through an orifice 1 mm in diameter. A manually operated vacuum pump is usually used.

These impingers are available in various kits containing several impingers, and in units providing considerable automation. Figure 2-35 shows a 24-port sequential air sampler designed for use with midget impingers and Greenburg-Smith impingers. It is designed to collect air-

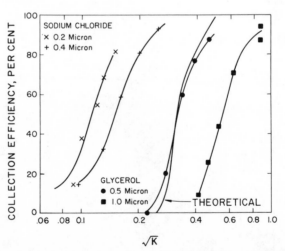

Figure 2-33. Collection efficiency of Greenburg-Smith impinger with sodium chloride and glycerol aerosols. Reprinted with permission from Schadt, C., and R. D. Cadle, *Anal. Chem.* **28**, 864 (1957). Copyright by the American Chemical Society.

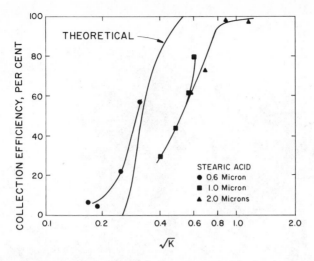

Figure 2-34. Collection efficiency of Greenburg-Smith impinger with stearic acid aerosols. Reprinted with permission from Schadt, C., and R. D. Cadle, *Anal. Chem.* **28**, 864 (1957). Copyright by the American Chemical Society.

borne contaminants from as many as 24 successive air samples over a period of up to 48 hr. In addition to automatically collecting 24 samples, the sampler can be programmed to collect 6, 12, or 18 discrete samples. A cycle timer controls the time from the start of one sample to the start of the next, and can be adjusted for 15-, 30-, 60-, or 120-min cycles. The free air flow rate can be regulated to 5 liters per minute maximum, and the lowest pressure attainable is 230 torr.

6. THERMAL PRECIPITATION

6.1. Theory

When a particle is suspended in a gas so that a temperature gradient exists across the particle, a force is exerted on the particle in the direction of decreasing temperature. Thermal precipitators operate on this principle. The aerosol from which particles are to be collected is drawn between two surfaces, one heated and the other usually at the aerosol temperature, and the particles move to the colder surface on which they are collected. The surfaces are placed as close together as is practical, because this maximizes the thermal gradient for any temperature difference between the surfaces.

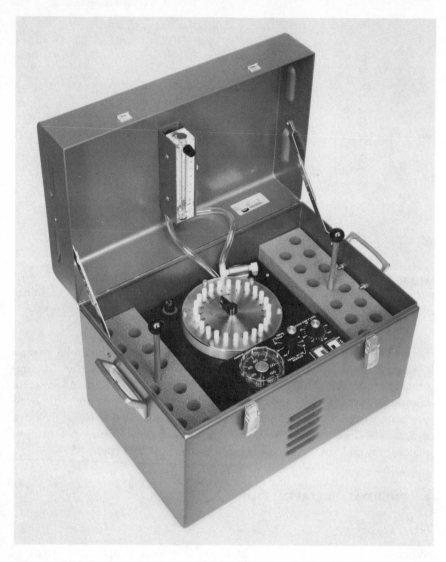

Figure 2-35. Sequential air sampler. Courtesy of The Bendix Corporation, Baltimore, Md.

This motion of a particle under the influence of a temperature gradient is called thermophoresis. The first description of thermophoresis was by Tyndall,[88] who observed a particle-free region surrounding a hot rod inserted in an aerosol. This region became evident as a dark space surrounding the rod when the aerosol was illuminated.

The theory of thermophoresis is quite different for particles that are small relative to the mean free path of the gas in which they are suspended (large Knudsen number) and for those that are large relative to the mean free path (small Knudsen number). These, of course, are the limiting situations.

The theory is relatively simple for large Knudsen numbers. The thermal conductivity of the particles is not important, and the thermal force results from the net impulse in the direction of the gradient imparted to the particle by the impinging gas molecules. The theory has been treated by Waldmann,[89] by Bakanov and Derjaguin,[90] and by Mason and Chapman.[91] They derived very similar equations. The Waldmann equation for the thermophoretic force is

$$F = \frac{32}{3} r^2 \cdot \frac{1}{5} \lambda_{\text{trans}} \frac{\Delta T}{\bar{c}} \tag{2-79}$$

where λ_{trans} is the translational part of the heat conductivity, ΔT is the temperature gradient, and $\bar{c}$ is the mean velocity of the molecules. The corresponding particle velocity is

$$v = \frac{1}{5(1 + \pi a/8)} \cdot \frac{\lambda_{\text{trans}}}{p} \Delta T \tag{2-80}$$

where a is the accommodation coefficient (about 0.85) and p is the pressure. This can be converted into the form

$$v = \frac{\bar{c}}{4(1 + \pi a/8)\, a} \frac{\lambda}{T} \Delta T \tag{2-81}$$

where λ is the mean free path of the gas molecules. The corresponding expression by Bakanov and Derjaguin is

$$v = \frac{15\pi}{6} \cdot \frac{\bar{c}}{8} \cdot \frac{\lambda}{T} \Delta T \tag{2-82}$$

The Mason and Chapman equation differs from Waldmann's in that the term $(1 + \pi a/8)$ is missing. It was derived on a somewhat different basis than the other equations. Mason and Chapman pointed out that very small particles can be considered to be large molecules; thus the thermophoresis of very small particles is the limiting case of molecular thermal diffusion in dilute gases.

The force on a particle having a small Knudsen number can be explained in terms of radiometer theory, which in turn is based on the concept of thermal creep developed by Maxwell. Any small portion of the surface in a thermal gradient is bombarded by more molecules from the low-temperature direction than from the high-temperature direction. However, all the molecules will leave the surface with essentially the same energy, assuming an accommodation coefficient of unity, with the result that gas will flow along the surface in the direction of increasing temperature. This flow causes both the rotation of the radiometer and the thermal forces on particles.

Epstein[92] derived an equation that was long accepted for large particles suspended in a gas in which a uniform thermal gradient exists at appreciable distances from the particles. The temperature distribution on the particle surface was calculated from the thermal conductivities of the particle and gas, assuming the applicability of the Fourier equation for heat conduction without convection. Epstein then applied the concept of thermal creep, obtaining an equation for gas flow along the surface. This equation was used as a boundary condition for solving the Navier–Stokes flow equations which yielded the following expression for the thermal force on a particle:

$$F = 9\pi r \left(\frac{k_a}{2k_a + k_i}\right) \frac{\eta^2}{\rho T}\, \Delta T \tag{2-83}$$

where k_a and k_i are the thermal conductivties of the gas and the particle, respectively.

Rosenblatt and LaMer[93] and later Saxton and Ranz[94] compared experimentally determined values for the thermal forces on particles with the predictions from Epstein's equation. They obtained good agreement, although all the studies were undertaken with particles having low thermal conductivities (paraffin oil, castor oil, and tricresyl phosphate). The experimental values were obtained with a modified Millikan oil drop apparatus which produced thermal gradients as well as electrical fields. Thermal forces were calculated from the settling rates of the particles in the presence and absence of thermal gradients. The particles were charged, and the electric fields were used to control the location of the particles in the Millikan cells.

As part of a comparison of the collection efficiencies of various commonly used aerosol particle sampling devices with the theoretical predictions of these efficiencies, Schadt and Cadle[95] determined the thermal forces on aerosol particles in a thermal precipitator under greatly varying conditions of thermal gradient, particle size, and thermal conductivity of the particles. The effect of thermal conductivity was especially interesting because the Epstein equation predicts that large aerosol particles having a high thermal conductivity would be very difficult to collect with a thermal precipitator. However, no such difficulties had been reported in the literature, although thermal precipitators had been used for many years for collecting airborne particles of many types.

The aerosol particles consisted of stearic acid, sodium chloride, or carbonyl iron, and each aerosol had a narrow particle size distribution. The thermal gradients were produced with a Casella thermal precipitator, using a ribbon rather than a wire for the hot surface. Thermal forces were calculated with the Stokes–Cunningham equation from the width of the deposit beneath the ribbon produced as the aerosol flowed between the ribbon and the cold surface in a direction normal to the length of the ribbon.

The experimentally determined thermal forces on stearic acid particles were nearly identical with those predicted by the Epstein equation, and the effects of particle size and the magnitude of the thermal gradient were close to those expected. However, the thermal forces on the highly thermally conducting sodium chloride and iron particles were 20–40 times the predicted values.

Because of the rather surprising nature of these results, they were confirmed using a modified oil-drop apparatus, and particles of tricresyl phosphate, sodium chloride, and mercury.[96] Two methods were used to measure the thermal forces, one involving measuring the difference in fall rate in the presence and absence of a thermal gradient and the other involving balancing the thermal and gravitational forces with the electric field so that the particle being observed remained stationary. Results obtained by the two methods agreed quite well and confirmed those obtained with the thermal precipitator. The thermal forces on the droplets of mercury were about 50 times the theoretical value.

Brock[97,98] developed an equation for the small Knudsen number case to explain the results of Schadt and Cadle. In addition, he and others experimentally confirmed these results. Brock's equation used the full set of Maxwell's classical boundary conditions, thermal creep, "temperature jump," and the friction slip. He resolved the steady-state Navier–Stokes equations for the case of a spherical particle suspended in a gas in the presence of a uniform temperature gradient and in the absence of external forces. He accounted for the effect of convective flow with a perturbation method.

Brock's equation is

$$F = 12\pi r^2 \eta \frac{C_{tm}(\lambda/r)}{(1 + 3C_m(\lambda/r))} \cdot \frac{k_a/k_i + C_t(\lambda/r)}{1 + 2(k_a/k_i) + 2C_t(\lambda/r)} \cdot \Delta T \quad (2\text{-}84)$$

where C_m is the tangential momentum first-order slip coefficient, C_t is the temperature jump first-order slip coefficient, and C_{tm} is the thermal creep first-order coefficient. The evaluation of these coefficients must be made semiempirically, using methods suggested by Brock. Brock stated that equation (2-84) is only strictly applicable for values of λ/r less than 0.25, but that extrapolation of the data of Schadt and Cadle to such values indicates agreement with equation (2-84) within 10–30%. This equation also agrees well with the thermal force data of Rosenblatt and LaMer[93] for tricresyl phosphate aerosols. Brock has stated that the agreement between the Epstein equation and experimentally determined values of thermal forces for particles of low thermal conductivity is merely circumstantial and has been misleading.

Waldmann and Schmitt[99] have described the derivation by Derjaguin and Bakanov[100,101] of an equation for the velocity resulting from thermal forces on large particles:

$$v = \frac{2}{3} \frac{8k_a + k_i + 2C_t(\lambda/r) \, k_i}{2k_a + k_i + 2C_t(\lambda/r) \, k_i} \cdot \frac{k_a}{5p} \, \Delta T \quad (2\text{-}85)$$

Derjaguin and Bakanov had concluded that near a surface the thermal creep coefficient is actually only 1/35 of Maxwell's original value, which, if true, rules out explaining the thermophoresis of large particles by thermal creep. Instead, they considered a new term for the heat flux within the gas in addition to the ordinary Fourier term. Waldmann and Schmitt stated that they were unable to understand some of the steps taken in these calculations, and that they remain of the opinion that thermophoresis for small Knudsen numbers has to be described as a surface phenomenon.

Brock[102,103] has also developed equations that predict the thermal forces in the transition region $0.2 \leq \mathrm{Kn} \leq \infty$. He developed the theory for large but finite values of Kn with the momentum transfer method as the starting point. A single spherical particle of radius r in an effectively infinite gas with a temperature gradient that is constant at large distances from the particle was assumed. The gas as a whole was assumed to be in mechanical equilibrium. Brock's equation for a monatomic gas is

$$F = F^* \{ 1 - [0.06 + 0.9\alpha + 0.28\alpha(1 - \alpha k_a/2k_i)] \, r/\lambda \} \quad (2\text{-}86)$$

where $F*$ is the thermal force in the "free molecule region" (Kn = ∞) and is given by equation (2-79); and α is the momentum accommodation coefficient. The same equation applies to polyatomic gases, except that α is replaced by β, the thermal accommodation coefficient.

Brock[103] compared his theory with some experimental thermal force data. If his theory were exact, a plot of τ − 0.06 versus $\alpha[1 + 3.11 (1 - \alpha k_a/2k_i)]$ should yield a straight line through the origin with a slope of 0.09. The τ is defined by the equation

$$F = F^* \left(\exp - \frac{\tau r}{\lambda} \right) \tag{2-87}$$

The slope of the best fit to the experimental data for monatomic gases was 0.105.

A very similar phenomenon is photophoresis, which is the motion occurring when an airborne particle is subjected to an intense beam of light. The movement may be in any direction relative to the direction of the beam, and it may be linear, or in the form of spirals, loops, and the like. It probably results from the conversion of absorbed light to heat by the particle and is generally believed to be a radiometer effect, similar to that for particles in a thermal gradient. It has not been used for particle collection, for obvious reasons.

6.2. Instrumentation

Perhaps the best-known design of a sampling head for a thermal precipitator is that described by Watson[104] in 1936. The aerosol to be sampled is drawn at 6–7 cm³ per minute between two disk-shaped microscope cover glasses, each held in place on the end of a fairly massive cylindrical block having a diameter equal to that of the glasses. These glasses are the cool surface on which the particles are collected, and are separated by a distance of 0.5 mm. A nichrome resistance wire, heated to 120°C by passing an electrical current through it, is stretched between the cover glasses. This head is mounted on a small water tank. By allowing water to flow from the bottom of the tank, the aerosol is drawn at the desired rate through the sampling head. The entire system can be mounted on a tripod.

A more modern form of this instrument, using the original design of sampling head, is manufactured by C. F. Casella & Co., Ltd., of London. The entire instrument is in a box having the dimensions 19 × 19 × 37 cm. The sampling head can be detached and means other than the water aspirator used to draw air between the cover glasses. Furthermore, the power supply can be detached from the sampling head and worn on the belt for convenience.

Casella also manufactures a long-period thermal precipitator designed for sampling in British coal mines, which can operate unattended during a working period in a mine. Suction is provided by a battery-driven piston pump.

Numerous thermal precipitators have been designed to deposit the particles on moving surfaces, and some of these have been commercially available. An example is the Numinco model MIC 501. A single fan draws air into the instrument both for cooling and for sampling. The particles can be collected directly on a glass cover slide, electron microscope grid, or other thin substrate, and the substrates are mounted on a rotating plate moved by a synchronous motor at the rate of one revolution every 24 hr. The deposit is spread into a ring, making it possible to determine the time at which particles were collected. The sampling rate, as is usual for thermal precipitators, is quite slow: 5–6 cm³ per minute. All the particles are deposited on a very small area, 1/64 in², and an entire cross section of the ring deposit can be seen in a low-power microscopic field. Figure 2-36 is a diagram of the instrument.

Joseph B. Ficklen, III, has for years manufactured several types of thermal precipitators. The Konisampler is a compact instrument for field use which operates on 12 watts from mains or batteries. The area of particle deposit can be varied according to the dust concentration and the duration of sampling. Three types of Strong–Ficklen thermal precipitators are manufactured. The oscillating precipitator spreads a particle deposit beneath a hot wire by making a complete cycle of oscillation once every 3 min. The continuous version moves the hot wire at a speed of 1 mm hr⁻¹ for up to 36 hr. The gravimetric instrument deposits the particles beneath a stationary hot plate over an area of 5 cm². The sampling rate of the gravimetric instrument is up to 500 cm³ min⁻¹, an unusually fast sampling rate for a thermal precipitator.

Wilson and Orr[105] have determined the factors that are important to the design of a thermal precipitator for use at high altitudes. Their investigation included determinations of the effectiveness of thermal precipitation under different conditions of operating pressure, particle size, precipitation plate separation distance, and aerosol throughput velocity. They developed data designed to be a practical guide for the design of a thermal precipitator for the high-altitude sampling of atmospheric particulate matter. Much of the data was obtained with an experimental precipitator.

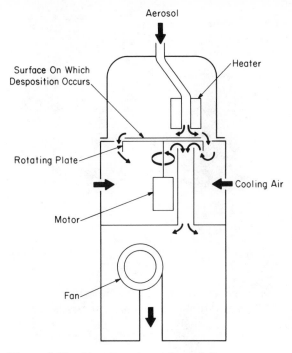

Figure 2-36.   Numinco thermal precipitator.

The precipitator they employed is shown in Figure 2-37. The aerosol entered the precipitator through an inlet in the center of the disk-shaped hot plate and flowed radially between the hot and cold plates. Particles were collected on a removable glass disk mounted on the cold plate.

The results were compared with theory and found to be generally consistent. They indicated an increase in collection efficiency, in terms of mass throughput capacity, as high as 700% for a decrease in aerosol pressure from 1.0–0.1 atm. They collected particles in the diameter range 2 μm to less than 30 Å at pressures as low as 2 torr (the Knudsen numbers were of the order of 10^4). Wilson and Orr concluded that for high-altitude sampling, a deposition area of about 0.3 m^2 would be needed for a sampling rate of 500 cfm at a pressure of 0.01 atm (an altitude of about 30 km), using a thermal precipitator operating at a temperature gradient of 2500°C cm^{-1}. They recommended a distance between precipitator plates of 1 mm, and thus an actual temperature difference of 250°C. They based this estimate on an increase in precipitator efficiency at 30 km of 2.5 times the efficiency at sea level, with an efficiency of essentially

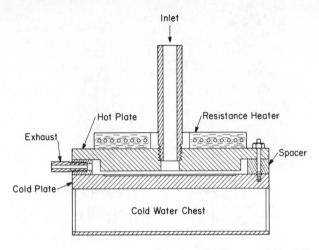

Figure 2-37.   Thermal precipitator of Wilson and Orr.[105]

100%, regardless of particle size. The literature describing various thermal precipitator designs and their performance is very extensive and has been reviewed by Gieseke.[106]

7.   ELECTROSTATIC PRECIPITATION

7.1.   Theory

Electrostatic precipitators are devices that utilize a corona discharge to produce a charge on the particles to be sampled, which in turn migrate to an electrode of opposite charge. A typical electrostatic precipitator consists of a cathode, usually a wire or set of points, that produces the discharge, an anode, often an aluminum cylinder concentric with the wire cathode, or a metal plate adjacent to the metallic point cathode, and some device for passing the aerosol between the anode and cathode. A high voltage is required to produce the corona discharge, usually between 5 and 15 kv. The precipitator need not consist of a single stage in which both charging and precipitation occur; two-stage precipitators charge the particles with a corona discharge in one stage and collect the particles in another, which usually has closer walls. The central wire or points can be the anode instead of the cathode, and alternating fields, corresponding to alternating currents, are sometimes used.

The theory of electrostatic precipitation has been reviewed by Rose and Wood[107] and by White,[108] among others. Two mechanisms are responsible for charging particles in the corona field of an electrostatic precipitator. The most important of these, usually called field or impact charging, is ion attachment. The other mechanism, diffusion charging, depends on the Brownian diffusion of the ions. According to White, field charging is especially important for particles exceeding 0.5 μm in diameter; the diffusion processes are especially important for particles smaller than 0.2 μm in diameter; both processes are important for particles of intermediate size. The theory of these processes has been developed primarily for industrial precipitators, but for the most part is applicable to air sampling devices.

The theory of field charging developed by White is based on theory developed by Rohmann[109] and by Pauthenier.[110] White assumes that all particles of the same size are charged to the same degree and that the particles are sufficiently far apart that interaction between them can be neglected. The electric field, E_0, and the gas-ion concentration, N_0, are assumed to be uniform in the neighborhood of each particle. The following equation is then derived for a conducting particle:

$$\frac{n}{n_s} = \frac{\pi N_0 e K t}{\pi N_0 e K t + 1} \tag{2-88}$$

where t is the time during which the particle has been exposed to the electric flux, K is the ion mobility, n is the particle charge, n_s is the saturation value of the particle charge, and e is the electric charge.

Since $\pi N_0 e K$ has the dimensions of reciprocal time, it can be denoted by

$$\frac{n}{n_s} = \frac{t}{t + t_0} \tag{2-89}$$

Since when $t = t_0$, $n/n_s = 1/2$, t_0 can be considered to be a time constant which indicates the charging rate. White gives as typical values $N_0 = 5 \times 10^8$ ions cm^{-3}, and $K = 2.2$ cm sec^{-1} volt^{-1} cm^{-1} = 600 esu of mobility. Thus since $e = 4.8 \times 10^{-10}$, $t_0 = 1/N_0 e K \pi = 0.0020$ sec. This is a very short time compared with the time an aerosol remains in the charging region of a electrostatic precipitator.

The maximum charge acquired by a particle is $n_s e$, given by the equation

$$n_s e = 3 E_0 r^2 \tag{2-90}$$

thus n_s is proportional to the field strength and to the surface area of the particle. This is an important consideration in the design of instruments that classify particles according to size on the basis of the mobility of the charged particles in an electric field. When the particles consist of dielectric rather than conducting materials, the only modification is in the equation for n_s:

$$n_s = \left(1 + 2\frac{k-1}{k-2}\right)\frac{E_0 r^2}{e} \tag{2-91}$$

where k is the dielectric constant.

Arendt and Kallman[111] derived the following equation for diffusion charging, the mechanism of importance for relatively small particles:

$$\frac{dn}{dt}\left(1 + \frac{r^2 C}{4nke}\right) = \pi r^2 C N_0 \exp{-\frac{ne^2}{rkT}} \tag{2-92}$$

where T is the absolute temperature and C is the root mean square velocity of the ions. White derived a similar equation which omits the term in brackets on the left side of equation (2-92):

$$\frac{dn}{dt} = \pi r^2 C N_0 \exp{\frac{-ne^2}{rkT}} \tag{2-93}$$

Equation 2-93 is readily integrated to yield

$$n = \frac{rkT}{e^2}\log\left(1 + \frac{\pi r C N_0 e^2 t}{kT}\right) \tag{2-94}$$

Equations for particles of intermediate size are very complicated and are usually solved by numerical methods using computers.

Numerous comparisons have been made of the theoretical predictions with experimental results. Rather good agreement was obtained when the particles were relatively large (>0.2 μm radius) and charging rates were low. However, Lippmann[33] quotes Mercer[112] as having investigated the charging characteristics of submicron particles and finding that the observed charges greatly exceeded those calculated from field and diffusion charging theory. Apparently the rate of charging of such small particles is faster than the theoretical diffusion charging rate, and a larger percentage of the ions reach and attach to such particles than the theory suggests, possibly because the electric field causes an increased ion concentration near the particles. On the other hand, Hewitt[113] using oil fume particles of 0.065–0.65 μm radius and ions of positive polarity found good agreement with diffusion-charging theory and quite good agreement with field-charging theory for those particle size ranges in which the theory would be expected to apply.

After the particles are charged they are collected by passing the aerosol through a continuation of the corona field in single-stage precipitators or by passing the aerosol through the field produced by the nondischarging plates of a second stage.

For fine particles the acceleration of gravity can be neglected, and the differential equation for motion in terms of acceleration is

$$m \, \frac{dv}{dt} = qE_C - 6\pi rv\eta \qquad (2\text{-}95)$$

where m is the particle mass, v is the velocity in the direction of the collection surface, E_C is the collection field and equals E_0 in single-stage collectors, and q is the charge. Integration of equation (2-95) yields

$$v = \frac{qE_C}{6\pi \eta r}\left(1 - \exp\frac{-6\pi \eta r}{m}\, t\right) \qquad (2\text{-}96)$$

The exponential term is very small because of the small value of m and can be ignored, yielding the equation

$$v = \frac{qE_C}{6\pi \eta r} \qquad (2\text{-}97)$$

Then from equation (2-90) for field charging,

$$v = \frac{3E_0 r^2 E_C}{6\pi \eta r} = \frac{E_0 E_C r}{2\pi \eta} \qquad (2\text{-}98)$$

or, for single-stage precipitators,

$$v = \frac{E_0^2 r}{2\pi \eta} \qquad (2\text{-}99)$$

These equations are based on the assumption that the air is a continuous medium and that there is no influence of a turbulent wake; in other words, that Stokes' law would apply. If the particles are small relative to the mean free path, the Cunningham correction must be applied; the equations would not be expected to yield useful results for Reynolds numbers much greater than unity. Also, of course, these equations only apply when the aerosol flow through the collection region is laminar.

Rose and Wood[107] derived the following equation for laminar flow through a tubular precipitator having an axial wire electrode:

$$x = 2\,\frac{Rv'}{v}\left[\frac{x'}{R} - \frac{1}{3}\left(\frac{x'}{R}\right)^3\right] \qquad (2\text{-}100)$$

where x is the distance along the wire, R is the radius of the tube, x' is the radial distance from the wire, and v' is the velocity of the aerosol flow. This equation is intended to take account of the effect of the parabolic aerosol velocity distribution across the tube. All the particles of a given size and thus the same transverse velocity, v, will be deposited within a tube length, L, given by the equation

$$L = \frac{4}{3} \frac{Rv'}{v} \tag{2-101}$$

Much of the theory of electrostatic precipitators has been developed for application to large-scale precipitators designed to collect industrial effluents. The airflow through such precipitators is usually turbulent. For such conditions Rose and Wood suggested the equation

$$E = 100 \left(1 - \exp\frac{-Av}{V} \right) \tag{2-102}$$

where E is the efficiency in percent, A is the total electrode collecting surface in square meters, and V is the volumetric flow rate in cubic meters per second. White[108] has applied this equation to various particle size distributions, such as normal and log normal distributions.

7.2.   Instrumentation

One of the oldest and best known of the electrostatic precipitators for aerosol sampling is that manufactured by the Mine Safety Appliance Co. (MSA). The sampling rate, 85 liters per minute, is kept constant by a voltage stabilizer in the power supply, which operates on power between 105 and 120 volts, 60-cycle ac. The precipitator is of the wire and cylinder variety, and the air is drawn through by a fan. The complete instrument in its carrying case weighs about 30 lbs.

In some countries such as England the power supply is 50 instead of 60 cycles, decreasing the flow rate through the MSA precipitator to about 70 liters per minute. Roach[114] developed a simple, rapid, null-point method for determining the actual flow rate through the MSA and similar electrostatic precipitators. The apparatus consisted of an air mover capable of delivering a flow rate in excess of that of the precipitator being tested, a vessel of about 20 liters capacity, and a manometer.

A high-volume electrostatic precipitator was designed by Lippmann et al.[115] and is manufactured by Del Electronics Corp. It is shown schematically in Figure 2-38, arranged for duct sampling. Unlike the MSA precipitator, this one has two stages, one for charging and the other for collection. The charging section has the common arrangement of an

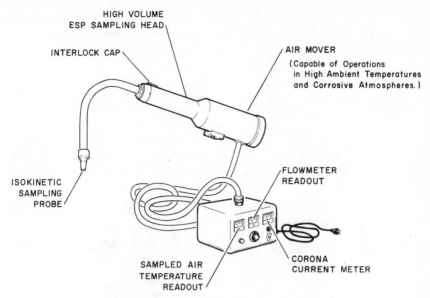

Figure 2-38.   High-volume ESP sampler for duct sampling.[115]

axial wire in a tube. However, the corona current is very high, permitting the unusually high sampling rate of 27 cfm. Some of the charged particles are collected on the cylinder walls within the charging zone. The collector zone reduces the interelectrode spacing by splitting the flow stream into two annuli, one inside and one outside the sampling tube. The outer stream is confined within an outer cylinder, and the particles are collected on both sides of the negatively charged sampling tube. This arrangement reduces the interelectrode spacing without reducing the free cross-sectional area. This instrument is equipped with a flowmeter consisting of a miniature bypass rotameter in the back cap of the sampler housing.

Schadt et al.[116] developed an electrostatic precipitator that continuously collects and measures particulate electrolytes suspended in air or other gases. A rotating disk anode is continuously washed with a stream of water or extremely dilute sulfuric acid. The electrical conductance of the resulting solution is measured by passing it through a conductivity cell, and recorded on a strip chart. The overall construction is shown in Figure 2-39, and the construction of the precipitation chamber and the electrodes is shown in Figure 2-40. Both the cathodes (points) and the rotating disk anode were made of stainless steel. The disk was rotated

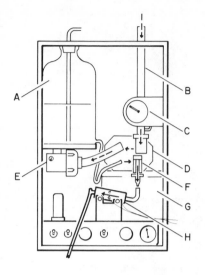

Figure 2-39. A schematic drawing of the precipitator and a flow diagram indicating the movement of air and of liquids. A indicates the reservoir bottle (2 liter capacity); B, sampling tube (5/8 in inside diameter); C, air flowmeter (modified airplane rate-of-climb meter); D, precipitation chamber (2 by 4 by 5 inches); E, blower (Oster no. 2); F, leveling device; G, power supply (6,000 to 10,000 volts D.C. current); H, conductivity cell (cell constant about 20 cm^{-1}). The arrow with broken shaft indicates the air path; that with unbroken shaft, the water path. From Schadt et al., *Arch. Ind. Hyg. Occ. Med.* **1**, 556 (1950). Copyright by American Medical Association.

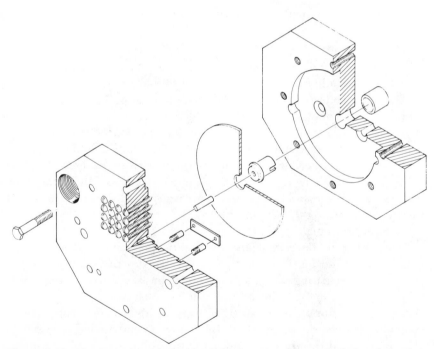

Figure 2-40. Construction of the precipitation chamber and of the electrodes. From Schadt et al., *Arch. Ind. Hyg. Occ. Med.* **1**, 556 (1950). Copyright by American Medical Association.

at one rpm by a synchronous motor. The chamber itself was fabricated from methyl methacrylate (Lucite). The liquid in the bottom of the chamber forms a bath through which the anode revolves, washing precipitated material from the anode. The washwater can, of course, be examined in a number of ways in addition to being drawn through a conductivity cell. For example, turbidity could be continuously measured to indicate the concentrations of insoluble airborne particles.

Another electrostatic precipitator that collects the particles in a liquid is manufactured by the Environmental Research Corporation (Figure 2-41). It consists of two basic components: a sampler module and a control module. The aerosol enters the sampler module through a calibrated nozzle and flows through the center of a fixed high-voltage plate. The aerosol then flows outward between this plate and a rotating collection disk at ground potential. The air is exhausted by a blower. A continuous negative corona is emitted by a ring of needles mounted on the high-voltage plate. Liquid flows in a thin film, supplied by a peristaltic pump, over the rotating collection disk, and the centrifugal force brings the liquid into a pickup dish. The liquid containing the particles is then pumped into a reservoir. Two models are available, one sampling air at 300–1,200 liters per minute and the other sampling at 2,500–15,000 liters per minute.

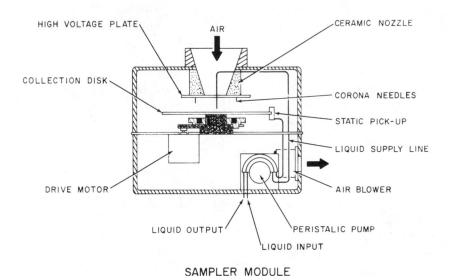

SAMPLER MODULE

Figure 2-41.   Environmental Research Corp. "LEAP" electrostatic precipitator.

A similar sampler is manufactured by Litton Systems, Inc. The liquid is spread over the collection plate surface with a monofilament wiper which is adjustable to achieve the best condition for various liquids. The collection surface and wiper can be observed through a Plexiglass window in the collection chamber.

Liu et al.[117] have described a compact, two-stage electrostatic precipitator manufactured by Thermo-Systems. They point out that in the conventional single-stage electrostatic precipitator, the collecting surface must also conduct the corona current and that a nonconducting surface could acquire sufficient electrical charge to repel the ions as well as the charged particles. Avoiding this difficulty is an advantage of the separate charging and precipitating regions. Size classification of the deposited aerosol, which occurs in most electrostatic precipitators, was minimized by using a pulsed precipitating electric field. Sampling efficiency and deposit uniformity were determined using monodispersed fluorescent aerosol particles (ranging in size from 0.028 to 3.2 μm diameter). The collection efficiency increased with increasing size from about 60 to 80%.

Morrow and Mercer[118] have built a point-to-plane electrostatic precipitator designed to collect airborne particles for study and measurement by electron microscopy. Basically, the instrument is a sharply-pointed electrode held in opposition to the flat end of a grounded rod. The space between is surrounded by an insulated duct through which the aerosol is drawn. The point electrode is a phonograph needle, and a standard electron microscope specimen holder is fitted to the upper portion of the rod so that the electron microscope grid covers the upper, flat end of the rod. A filter holder and filter back up the precipitator. The instrument body is constructed of methacrylate plastic, the needle electrode is removable, all threaded connections are sealed with O-rings, and the position of the rod-electrode is adjustable. The length of the duct, from the inlet to the filter holder, is about 8 cm. The potential of the point electrode is normally negative. The usual aerosol flow rate is about 70 cm³ min⁻¹.

Several comparisons were made of the size distributions of particles collected with this precipitator and of particles collected from the same aerosol with various thermal precipitators. The results are shown in Table 2-3. Quite satisfactory agreement was obtained between the two methods for several different types of particulate material.

Adley[119] has described a reciprocating electrostatic precipitator for the collection of specimens for particle size determination. Both the theory of electrostatic precipitation and experimental data demonstrate that the particle size distribution of the deposited particles changes with dis-

Table 2-3.   Particle Size Analyses of Dusts.[118]

Exp.	Aerosol	Point to Plane		Thermal		
		GMD	σ_g	GMD	σ_g	
10–62	UO_2	0.39	2.25	0.39	2.34	Casella
				0.32	2.03	Oscillating
6–19	UO_2	0.36	1.80	0.35	1.82	Oscillating
				0.41	1.85	Casella
516	Fe_2O_3	0.15	1.89	0.16	1.87	Casella
ND 63	CrO	0.07	1.40	0.06	1.42	Walkenhorst

tance from the region of initial charging. Therefore particles should be measured and counted on many different, representative parts of the collection surface. Adley's instrument was designed to minimize such bias. A bar 6 in. long fits into a longitudinal slot in the collection tube of a wire and tube type precipitator so that the bar and tube are flush. A cylindrical cam causes the bar to reciprocate once each minute. Electron microscope grids are mounted on the bar so that they completely traverse the collection zone.

A problem sometimes encountered when using electrostatic precipitators for sampling results from the fact that they partially convert oxygen to ozone which may attack and destroy organic particles.

REFERENCES

1. Chen, C. Y., *Chem. Rev.* **55**, 595 (1955).
2. Dorman, R. G., in E. G. Richardson, Ed., *Aerodynamic Capture of Particles,* Pergamon, New York, 1960.
3. Dorman, R. G., in C. N. Davies, Ed., *Aerosol Science,* Academic Press, New York, 1966.
4. Pich, J., ibid.
5. Davies, C. N., *Air Filtration,* Academic Press, London, New York, 1973.
6. Lamb, H., *Hydrodynamics,* 6th ed., Cambridge University Press, London, 1932.
7. Landahl, H., and K. Herrmann, *J. Colloid Sci.,* **4**, 103 (1949).
8. Ranz, W. E., and J. B. Wong, *Ind. Eng. Chem.,* **44**, 1371 (1952).
9. Albrecht, F., *Physik. Z.,* **32**, 48 (1931).
10. Sell, W., *Forsch. Gebiete Ingeniurw.* **2**, Forschungsheft, 347 (August, 1931).

11. Langmuir, I., and K. B. Blodgett, General Electric Research Laboratory, Schenectady, N.Y., Report RL 225, 1944–45.

12. Davies, C. N., and C. V. Peetz, *Proc. Roy. Soc.*, **A234**, 269 (1956).

13. Sinclair, D., Handbook on Aerosols, U.S. Atomic Energy Commission, Washington, D.C. 1950.

14. Gregory, P. H., *Ann. Appl. Biol.*, **38**, 357 (1951).

15. Wong, J. B., W. E. Ranz, and H. F. Johnstone, *J. Appl. Phys.*, **26**, 244 (1955).

16. Fuchs, N. A., *The Mechanics of Aerosols*, Pergamon, Oxford, 1964.

17. Langmuir, I., OSRD Report No. 865, Office of Techanical Services, Washington, D.C., 1942.

18. Ranz, W. E., *Engng. Exp. Sta. Rept. No. 8*, University of Illinois, Urbana, 1953.

19. Davies, C. N., *Proc. Inst. Mech. Engrs. (London) B1*, 185 (1952).

20. Ramskill, E. A., and W. L. Anderson, *J. Colloid Sci.*, **6**, 416 (1951).

21. Stern, S. C., H. W. Zeller, and A. C. Schekman, *J. Colloid Sci.*, **15**, 546 (1960).

22. Cadle, R. D., *EOS*, **53**, 812 (1972).

23. Friedlander, S. K., *J. Colloid Interface Sci.*, **23**, 157 (1967).

24. Spielman, L., and S. L. Goren, *Environ. Sci. Tech.*, **2**, 279 (1968).

25. Cadle, R. D., and W. C. Thuman, *Ind. Eng. Chem.*, **52**, 316 (1960).

26. Sullivan, R. R., and K. L. Hertel, *Advances in Colloid Science*, Vol. 1, Interscience, New York, 1942.

27. Brinkman, H. C., *Appl. Sci. Res. A1*, **27**, 81 (1947).

28. Spurny, K. R., J. P. Lodge, Jr., E. R. Frank, and D. C. Sheesley, *Environ. Sci. Tech.*, **3**, 453 (1969).

29. Pich, J., *Staub*, **24**, 2 (1964).

30. Spurny, K., and J. Pich, *Int. J. Air Wat. Poll.*, **8**, 193 (1964).

31. Megaw, W. J., and R. D. Wiffen, *Inter. J. Air. Wat. Poll.*, **7**, 501 (1963).

32. Walkenhorst, W., *Staub*, **19**, 69 (1959).

33. *American Conference of Governmental Industrial Hygienists, Air Sampling Instruments*, 3rd ed., American Conference of Governmental Industrial Hygienists, 1014 Broadway, Cincinnati, Ohio, 1966.

34. Green, H. L., British Patent 432, 137; 1936.

35. Green, H. L., and Thomas, D. J., British Patent 727, 975; 1955.

36. Stafford, E., and W. S. Smith, *Ind. Eng. Chem.*, **43**, 1346 (1951).

37. First, M. W., J. B. Graham, G. M. Butler, C. B. Walworth, and R. P. Warren, *Ind. Eng. Chem.*, **48**, 696 (1956).

38. First, M. W., and J. B. Graham, *Ind. Eng. Chem.*, **50**, 63A (1958).

39. Pate, J. B., and R. D. Cadle, in W. H. Mathews, W. W. Kellogg, and G. D. Robinson, Eds., *Man's Impact on the Climate,* MIT Press, Cambridge, Mass., 1971; K. R. Spurny, J. P. Lodge, Jr., E. R. Frank, and D. C. Sheesley, *Environ. Sci. Technol.,* 3, 464 (1969).

40. Lodge, J. P., Jr., J. B. Pate, B. E. Ammons, and G. A. Swanson, *J. Air Poll. Cont. Assoc.,* 16, 197 (1966).

41. Whytlaw-Gray, R., and H. S. Patterson, *Smoke: A Study of Aerial Disperse Systems,* Edward Arnold & Co., London, 1932.

42. Millikan, R. A., *Phys. Rev.,* 15, 545 (1920).

43. Millikan, R. A., *Phys. Rev.,* 21, 217 (1923); 22, 1 (1923).

44. Knudsen, M., and S. Weber, *Ann. Phys. Lpz.,* 36, 982 (1911).

45. Mattauch, J., *Z. Phys.,* 32, 439 (1925).

46. Davies, C. N., in "Symposium on Particle Size Analysis," Institution of Chemical Engineers, London, February 4, 1947.

47. Green, H. L., and W. R. Lane, *Particulate Clouds: Dusts, Smokes, and Mists,* 2nd ed., Van Nostrand, New York, 1964.

48. Schadt, C., and R. D. Cadle, *Anal. Chem.,* 29, 864 (1957).

49. Cadle, R. D., and P. L. Magill, *Ind. Eng. Chem.,* 43, 1331 (1951).

50. Schaefer, V. J., in W. H. Mathews, W. W. Kellogg, and G. D. Robinson, Eds., *Man's Impact on the Climate,* MIT Press, Cambridge, Mass., 1971.

51. Davies, C. N., Ed., *Aerosol Science,* Academic Press, New York, 1966.

52. Jacobs, M. B., *The Chemical Analysis of Air Pollutants,* Interscience, New York, 1960.

53. Lane, B. W., and P. E. Clark, Jr., *Air Water Int. J.,* 10, 627 (1966).

54. Lucas, D. H., *Atmos. Environ.,* 1, 71 (1967).

55. Kellogg, W. W., R. D. Cadle, E. R. Allen, A. L. Lazrus, and E. A. Martell, *Science,* 175, 587 (1972).

56. Lodge, J. P., Jr., J. B. Pate, W. Basbergill, G. S. Swanson, K. C. Hill, E. Lorange, and A. L. Lazrus, Chemistry of United States Precipitation, National Center for Atmospheric Research, Boulder, Colorado, 1968.

57. Dautrebande, L., Private communication.

58. Fuchs, N. A., and I. Petrjanoff, *Nature,* 139, 111 (1937).

59. May, K. R., *J. Sci. Instr.,* 22, 187 (1945).

60. May, K. R., *J. Sci. Instr.,* 27, 128 (1950).

61. Farlow, N. H., *J. Colloid Sci.,* 11, 184 (1956).

62. Schaefer, V. J., The Preparation and Use of Water Sensitive Coatings for Sampling Cloud Particles, General Electric Co. AAF Technical Report 5539, 1947.

63. Farlow, N. H., and F. A. French, *J. Colloid Sci.,* 11, 177 (1956).

64. May, K. R., *Nature,* 183, 742 (1959).

65. Pollak, L. W., *Int. J. Air Poll.,* **1,** 293 (1959).

66. Mason, B. J., *The Physics of Clouds,* 2nd ed., The Clarendon Press, Oxford, 1971.

67. Ranz, W. E., Principles of Inertial Impaction, Bulletin No. 66, Dept. of Engineering Research, The Pennsylvania State University, University Park, Pa., 1956.

68. Ranz, W. E., and J. B. Wong, *A.M.A. Arch. of Indus. Hyg. Occ. Med.,* **5,** 464 (1952).

69. Ranz, W. E., Personal communication.

70. Rosin, P., E. Rammler, and W. Intelmann, *Z. Ver. Deut. Ing.,* **76,** 433 (1932).

71. Magill, P. L., M. Rolston, J. A. MacLeod, and R. D. Cadle, *Anal. Chem.,* **22,** 1174 (1950).

72. Johnstone, H. F., R. B. Feild, and M. C. Tassler, *Ind. Eng. Chem.,* **46,** 1601 (1954).

73. Wolff, H. S., and S. A. Roach, in *Inhaled Particles and Vapours,* C. N. Davies, Ed., Pergamon, London, 1961.

74. Kanter, C. V., R. G. Lunche, and A. P. Fudurich, *J. Air. Poll. Control Assoc.,* **6,** 191 (1957).

75. Stevens, D. C., and J. Stephenson, *J. Aerosol Sci.,* **3,** 15 (1972).

76. Task Group on Lung Dynamics, *Health Physics,* **12,** 173 (1966).

77. Harris, G. W., and B. A. Maguire, *Ann. Occup. Hyg.,* **11,** 195 (1968).

78. Stevens, D. C., and W. L. Churchill, *J. Aerosol Science,* **4,** 85 (1973).

79. Marple, V. A., and B. Y. H. Liu, *Environ. Sci. Technol.,* **8,** 648 (1974).

80. Gallily, I., and V. K. LaMer, *J. Phys. Chem.,* **62,** 1295 (1958).

81. Dahneke, B., *J. Coll. Interface Sci.,* **37,** 342 (1971).

82. Jordon, B. W., *Brit. J. Appl. Phys. Suppl.,* **3,** 5194 (1954).

83. Dessens, H., *Quart. J. Roy. Met. Soc.,* **75,** 23 (1949).

84. Rinehart, G. S., *J. Col. Interface Sci.,* **44,** 546 (1973).

85. Junge, C. E., and J. E. Manson, *J. Geophys. Res.,* **66,** 2163 (1961).

86. Greenburg, L., and G. W. Smith, *U.S. Bur. Mines Rept. Invest.,* No. 2392 (1922).

87. Littlefield, J. B., F. L. Feicht, and H. H. Schrenk, *U. S. Bur. Mines Invest.,* 3360 (1937).

88. Tyndall, J., *Proc. R. Instn. G. Br.,* **6,** 3 (1870).

89. Waldmann, L., *Z. Naturf.* 14a, 589 (1959).

90. Bakanov, S. P., and B. V. Derjaguin, *Dis. Faraday Soc.,* **30,** 130 (1961).

91. Mason, E. A., and S. Chapman, *J. Chem. Phys.,* **36,** 627 (1962).

92. Epstein, P. S., *Z. Physik,* **54,** 537 (1929).

93. Rosenblatt, P., and V. K. LaMer, *Phys. Rev.,* **70,** 385 (1946).

94. Saxton, R. L., and W. E. Ranz, *J. Appl. Phys.*, **23**, 917 (1952).

95. Schadt, C. F., and R. D. Cadle, *J. Coll. Sci.*, **12**, 356 (1957).

96. Schadt, C. F., and R. D. Cadle, *J. Phys. Chem.*, **65**, 1689 (1961).

97. Brock, J. R., *J. Coll. Sci.*, **17**, 768 (1962).

98. Brock, J. R., *J. Phys. Chem.*, **66**, 1763 (1962).

99. Waldmann, L., and K. H. Schmitt, in Aerosol Science, C. N. Davies, ed., Academic Press, New York, 1966.

100. Derjaguin, B. V., and S. P. Bakanov, *Dokl. Akad. Nauk SSSR (Phys. Chem.)* **141**, 384 (1961).

101. Derjaguin, B. V., and Y. I. Yalamov, *Dokl. Akad. Nauk SSSR (Phys. Chem.)* **155**, 886 (1964).

102. Brock, J. R., *J. Colloid Interface Sci.*, **23**, 448 (1967).

103. Brock, J. R., *J. Collid Interface Sci.*, **25**, 392 (1967).

104. Watson, H. H. *Trans. Inst. Min. Mett.*, **46**, 176 (1936).

105. Wilson, T. W., and C. Orr, Jr., An investigation of factors important in the design of a thermal precipitator for use at high altitude, Final Report, Project No. B-167, Engineering Experiment Station, Georgia Institute of Technology, Atlanta, Ga., 1962.

106. Gieseke, J. A., in W. Strauss, Ed., *Air Pollution Control, Part II*, Wiley-Interscience, 1972.

107. Rose, H. E., and A. J. Wood, *An Introduction to Electrostatic Precipitation in Theory and Practice*, Constable, London, 1956.

108. White, H. J., *Industrial Electrostatic Precipitation*, Wesley Publishing Co., Palo Alto, Calif., 1963.

109. Rohmann, H., *Z. Phys.*, **17**, 253 (1923).

110. Pauthenier, M. M., and M. Moreau-Hanot, *J. Phys. Rad.*, **3**, 590 (1932).

111. Arendt, P., and H. Kallman, *Z. Phys.*, **35**, 421 (1926).

112. Mercer, T. T., Charging and precipitation characteristics of submicron particles in the Rohmann electrostatic particle separator, UR 475, U. of Rochester Atomic Energy Project, Feb., 1957.

113. Hewitt, G. W., *AIAA Trans.*, **76**, 300 (1957).

114. Roach, S. A., *Am. Ind. Hyg. Assoc. J.*, **27**, 135 (1966).

115. Lippmann, M., H. J. Di Giovanni, S. Cravitt, and P. Lilienfeld, *Ind. Hyg. Assoc. J.*, **26**, 485 (1965).

116. Schadt, C., P. L. Magill, R. D. Cadle, and L. Ney, *Arch. Ind. Hyg. Occ. Med.*, **1**, 556 (1950).

117. Liu, B. Y. H., K. T. Whitby, and H. H. S. Yu, *Rev. Sci. Inst.*, **38**, 100 (1967).

118. Morrow, P. E., and T. T. Mercer, *Am. Ind. Hyg. Assoc. J.*, **25**, 8 (1964).

119. Adley, F. E., *Am. Ind. Hyg. Assoc. J.*, **19**, 75 (1958).

3

THE MEASUREMENT OF
COLLECTED PARTICLES

Once particles have been collected from an aerosol, many methods are available to measure them, either individually or *in toto*. The methods that can be used depend on the particle size (or size distribution), the nature of the surface on which the particles are collected, and to some extent the nature of the particles. For example, optical microscopy cannot be used to make measurements on particles much smaller than the wavelength of light. The electron beam of an electron microscope will destroy some types of particles, and "shadowing" the particles with the condensed vapors of some heavy metal such as gold, or preparing stable replicas of the particles which are then measured with the electron microscope, may be required. As mentioned earlier, particles collected on fiber filters are quite difficult to measure except for the weight of collected particulate material and the amount and chemical composition of extractable material. Some types of particles, especially relatively large solid particles, can be transferred from such filters into a liquid, for example by ultrasonic agitation, and the resulting suspension evaporated, depositing the particles onto a flat surface. Thus it is obvious that the collection method should be selected keeping in mind the available measurement techniques.

Sometimes it is helpful, or necessary, to use two or more methods of measurement. For example, if the size distribution is very broad a combination of optical and electron microscopy may be required to obtain

138

a reliable value for the particle size distribution. When using methods that yield a mean size, the use of two or more methods that are based on quite different scientific principles can often furnish at least some information concerning the distribution of sizes (light scattering is one such technique and determining the number of particles per unit weight of particulate material is another).

The methods described in detail in this chapter are those that are especially appropriate when very small masses and volumes of particulate material have been collected from an aerosol, as is usually the situation.

The techniques available when large amounts of particulate material are collected are, of course, much more numerous than when the amounts collected are very small; they are described in detail in a number of books devoted to the measurement of particle size distribution.[1-4] Some of these are mentioned briefly at the end of this chapter.

1. OPTICAL MICROSCOPY

Optical microscopy is one of the oldest methods for determining the sizes and shapes of fine particles and is still one of the preferable methods under certain conditions. Measuring a large number of individual particles is tedious unless some type of automatic scanning instrumentation can be used. Therefore, when large amounts of sample are available, much more rapid techniques than conventional microscopy are often appropriate. However, when only very small amounts of particulate material are available for making the desired measurements, for example, when particles are collected on a microscope slide by techniques such as thermal precipitation or sedimentation, microscopy may be the only available technique for obtaining the information desired.

As pointed out by Chamot and Mason,[5] in chemical technology, including particle measurement and identification, the microscope is used for two kinds of studies, namely, the revelation of fine structure and the determination of optical and other properties. Particle measurement uses the first of these, but the second may be used for identifying particles of a certain substance to be measured in a mixture of particles.

1.1. Theory

1.1.1. The Lens System

The most important part of a microscope is its basic lens system, for it is this that determines the resolution that can be achieved. Auxiliary lens systems are often also used, for example, for the illumination system and for photomicrography. The associated hardware can also be very important, especially when quantitative measurements are to be made. Not only should the stand be rigid; the various fine adjustments and calibrations should be carefully engineered and constructed. Nonetheless, a microscope can be no better than its lens system.

Figure 3-1 is a diagram of the optical system of a compound microscope and its illuminating system arranged to provide Köhler illumination (described later in this chapter). In addition to lenses, diaphragms and the lamp filament are shown. In its simplest form, the lens system

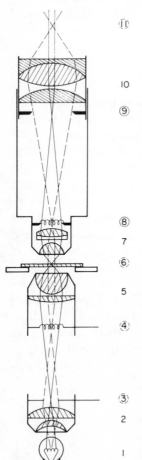

Figure 3-1.   Diagram of the optical systems of a compound microscope and an illuminating system arranged for Köhler illumination.[6] Courtesy of Karl Zeiss Inc., New York. The interlocking ray cones in a compound microscope: 1. Minimum size of the luminous area; 2. Lamp condenser; 3. Lamp condenser diaphragm = luminous field diaphragm; 4. Condenser iris = aperture diaphragm; 5. Condenser; 6. Illuminated specimen field; 7. Microscope objective; 8. Intermediate image of diaphragm 4; 9. Eyepiece diaphragm = field diaphragm; 10. Eyepiece; 11. Exit pupil of the microscope = position of the eye. For greater clarity, the figure is not drawn to scale. The various apexes of the cones are marked by circles. At all the points, 1, 3, 4, 6, 8, 9, and 11, the product of the area of the beam section and the solid angle of the cone apex is a constant (the law of constant light gathering power in the beam path of optical instruments with uninterrupted imaging). The light output in watts or photometrically in Lumens, is directly proportional to this constant product. The proportionality factor is called luminance (the law of constant luminance in the beam path in the absence of light losses due to absorption [no light filter] or reflection [coated optics]).

of a compound microscope consists only of two lenses: the objective and the eyepiece, which together with the eye produces an image on the retina. In the diagram two ray paths of light are shown, the solid line being the image-forming ray path and the broken line being the illuminating ray path. The figure obviously is not drawn to scale. When the microscope is focused on the object, at 6, the object is just beyond the focal plane of the objective, 7, and a real inverted image of the object is formed at 9. This real image must be formed at or very close to the focal plane of the eyepiece, 10, so that the light leaving the eyepiece is parallel or nearly so and can readily be focused onto the retina by the lens of the eye. Put another way, the eye is essentially focused for infinity when examining an object with the aid of a microscope, and since eyes vary considerably, an instrument that is focused for one person may be considerably out of focus for another.

The magnification Mo produced by the objective, defined as the length of the image over the length of the object, can be calculated from the equation.

$$Mo = \frac{\text{distance between image and back focal plane of objective}}{\text{focal length of objective}} \tag{3-1}$$

$$= \frac{\text{distance of the image from the ``effective center'' of the objective}}{\text{distance of the object from the ``effective center'' of the objective}}$$

In principle, we can achieve as great a magnification as desired by making the image distance sufficiently large or the focal length sufficiently small. In practice, the focal length is made shorter to increase the magnification, but as discussed later magnification beyond a certain degree is "empty magnification" and does not reveal more detail concerning the object observed. The magnification produced by an objective, usually based on a tube length of 160 mm, is commonly marked on the objective housing by the manufacturer.

When the light from the eyepiece is received, the eye forms a virtual image of the real image produced by the objective. This virtual image is, of course, the real image on the retina mentioned above and will appear to the observer to be at various distances, depending in part on how close to parallel is the light from the eyepiece and in part on the eye. The apparent size of the virtual image depends on the size of the image on the retina, and also on the apparent distance of the virtual image from the eye. However, it is desirable to be able to make some standard estimate of the amount of magnification provided by an eyepiece, if for no reason other than to compare eyepieces. For this purpose, the apparent size of the virtual image at 250 mm is used, and the magnification, *Me,* defined as length of the virtual image/length of the real image is obtained from the equation

$$Me = \frac{250}{\text{focal length of the eyepiece}} \qquad (3\text{-}2)$$

The approximate total magnification is then simply $Mo \times Me$. As Chamot and Mason[5] point out, although this expression is mathematically accurate, it is at best an approximation in practice because the focal lengths provided by manufacturers are usually given only to one or two significant figures; in addition, the tube lengths, indicated by graduations on some instruments, do not always correspond to the optical tube length. The measurement of fine particles by microscopy requires much more accurate means of determining the magnification, and these are discussed later in this chapter.

1.1.2. Objectives

Although the lens system is the most important part of the microscope, the objective is the most important part of the lens system (Figure 3-2). Three types of objectives are normally commercially available: achromats, semi-apochromats, and apochromats, listed in order of increasing excellence and complexity. They also vary markedly with regard to resolving power, which is the ability of the objective to reveal detail. The limitations of resolution are demonstrated in somewhat oversimplified fashion in Figure 3-3. The A and B represent dots or parallel lines that are close together and are observed with a microscope. Because of the limited resolving power of the objective, the images are somewhat wider than the dots or lines, and a plot of image density along an imaginary line through the dots or perpendicular to the two lines is somewhat like that in Figure 3-3a. The images somewhat overlap, but there is a deep enough "well" between them that the observer recognizes them as two objects. However, if the lines are closer together, as in Figure 3-3b, the observer sees a single image and the objects are not resolved. The vertical lines at A, B, A', and B' represent the abrupt discontinuities in the plots if the resolution were nearly perfect.

Most illuminated objects scatter or emit light in many directions, and the greater the range of directions from which the objective can gather light the greater the resolving power of the objective. This range is shown in Figure 3-1 as the one defined by the solid lines between 6 and 7. Since the object is located just outside the lower focal plane of the objective, the greater the focal length of the objective the wider must be

Figure 3-2. Several objectives having different numerical apertures. The objective on the left is for oil immersion. Courtesy of the American Optical Co.

the objective to attain the same angle of the cone of rays from the object. This is a reason in addition to that mentioned above for using very short focal lengths for microscope objectives, especially for those designed to have high resolving power. This corresponds to a very small "working distance" between the object and the front of the objective. Special objectives with greater than usual working distances have been designed for special purposes, for example, when the object must be observed through the top window of a cell.

The angle between the most divergent rays of light from the object that can pass through the front of the objective and form an image is termed the angular aperture (A.A.). However, although the angular aperture is important, the lowest refractive index between the object and the objective, n (or rather the real part of the complex refractive

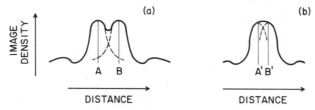

Figure 3-3. Diagramatic representation of resolving power. A, B, A′, and B′ are dots or parallel lines that are close together and are observed with a microscope.

index, $n + ik$), must also be considered. This was done by Ernst Abbe who defined the term numerical aperture, N.A., as:

$$N.A. = n \left(\sin \frac{A.A.}{2} \right) \qquad (3\text{-}3)$$

It is an excellent measure of resolving power and is marked on most microscope objectives along with the focal length and the magnification.

The maximum A.A. possible in principle is 180° [sin(A.A./2) = 1], and objective lens systems having values of sin (A.A./2) as high as about 0.95 have been constructed, although for achromats the upper limit is about 0.65. Since air has a refractive index close to unity, the numerical aperture of an objective can be increased markedly by eliminating air between the object and the objective using a cover slip and "immersion oil" in which the front of the objective is immersed. In this way numerical apertures as high as 1.4 have been achieved (Figure 3-4). The high numerical aperture of such objectives is only achieved when the object (the particles) is mounted in some medium such as cedar wood oil having a refractive index at least is high as that of the cover glass. Allen[7] states that if the object is mounted dry, so that air is between it and the cover glass, the effective N.A. will be at most 1.00. Objectives designed to be used dry are not satisfactory for use with an immersion oil. This is because the corrections for aberrations discussed below must be different for the different environments.

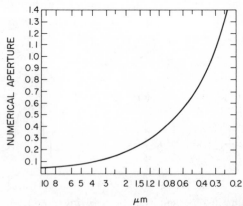

Figure 3-4. Relationship between resolving power in μm for green light ($\lambda =$ 550 nm) and numerical aperture. Courtesy of Ernst Leitz, Inc.

The smallest detail an objective can resolve was shown by Abbe to be given by the expression $\lambda/2\text{N.A.}$ where λ is the wavelength of light. When essentially white light, with a wide wavelength range, is used, a value of 0.5 μm serves as a fairly good representation of λ. Thus the resolving power of a microscope can be increased by decreasing the wavelength of the illuminating light. Blue light may be used for visual observation, but to obtain a marked increase in resolution by decreasing the wavelength ultraviolet light must be used. Then a real image rather than a virtual image must be produced by the eyepiece on a fluorescent screen or photographic emulsion. Television image scanning has also been used to observe the real images. Glass lenses transmitting wavelengths as short as 365 mμ have been prepared, but quartz lens systems that transmit down to about 275 mμ are available. Below 200 mμ air absorption of the radiation occurs and this region is called the vacuum ultraviolet region.

The ultraviolet microscope is still used, but to a considerable extent it has been superceded by the electron microscope, which has a very small numerical aperture and achieves its very high resolving power from the fact that the wavelength of the electrons is orders of magnitude shorter than that of visible light.

Abbe's theory of resolution is based on the theory of the diffraction of light by the parallel lines of a diffraction grating, and according to his theory if the direct ray and one of the diffracted rays is collected by the objective, the lines should be resolved. By using oblique illumination, both rays may be collected when otherwise only the direct ray would enter the objective, as shown by Figure 3-5. According to Chamot and Mason,[5] the resolving power of an objective can be nearly doubled by using oblique illumination.

The Abbe theory of resolution has not been universally accepted, and other theories have been developed. For one thing, Abbe's theory is based on the theory of diffraction gratings, and most objects observed with a microscope have a much more complex structure. Nonetheless, it has served very well as a basis for comparing the resolving power of

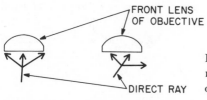

Figure 3-5. The use of oblique illumination to include rays diffracted by the object.

various objectives used under different conditions and to explain such aspects of microscopy as increasing resolution by using immersion lenses and by the use of oblique illumination. A theory developed by Rayleigh and by Helmholtz considers the image of a point. The image consists of a somewhat wider point surrounded by concentric rings decreasing in intensity with distance from the point (Figure 3-3). The diameter of the image point is given by the equation $1.2\lambda/2$N.A., and two adjacent image points will be resolved if the corresponding object points are this far apart. This expression is, of course, nearly identical with that given by Abbe.

There is no point to using magnifications greater than will readily reveal resolved details. Magnifications exceeding $1000\times$ are seldom useful, and the excess above that required to reveal resolved detail is often called empty magnification.

The numerical aperture of an objective also has a bearing on the depth of field. In principle, only an infinitely thin plane at the object will be in focus for a given microscope setting. If a point object is above or below this plane, a "circle of confusion" is produced; a small circle of confusion will not be noticed and will be acceptable. The total distance above and below the plane within which the circle of confusion is acceptable is the depth of field (sometimes called the depth of focus). The depth of field is obviously somewhat subjective, but it decreases with increasing numerical aperture until, for large numerical apertures, it is a fraction of a micrometer.

The depth of field can be calculated from the formula $c/\tan$ (A.A./2) where c is the diameter of the circle of confusion one is willing to accept. The angular aperture can, of course, be calculated from the numerical aperture marked on the objective using the formula $2 \sin^{-1}$ N.A. The depths of field for several numerical apertures are given in Table 3-1, based on an acceptable circle of confusion equal to the theoretical limit of resolution of the objective.

The limited depth of field can be useful for some purposes, such as measuring the thickness of particles, but it is often a nuisance when observing fine particles, because only a portion of a particle may be observed at one time if the particle is large relative to the depth of field.

D. McLachlan, Jr., has developed a clever photomicrographic method for circumventing the depth-of-field difficulty. The object is vertically scanned simultaneously with the plane in focus and with a plane of light that illuminates only the portion of the object in focus while exposing the photographic plate or film.

Table 3-1.   Properties of Objectives

Numerical Aperture (N.A.) of Objectives	Theoretical Limit of Resolution,* μm	Depth of Field,† μm
0.10	2.8	28
0.15	1.8	12
0.25	1.1	4.2
0.50	0.6	1.0
0.85	0.3	0.2

*Calculated from the formula $\lambda/2$ N.A., where λ is the wave length of the illumination and is assumed to be 0.55 μm.

†Calculated from the formula $c/\tan$ (A A./2), where c is the diameter of the "circle of confusion" which one is willing to accept and has been chosen to equal the limit of resolution, and A.A. is the angular aperture. The angular aperture can be calculated from the numerical aperture using the formula $2 \sin^{-1}$ N.A.

In visual microscopy a small depth of field is not quite so objectionable as in photomicrography. If a particle is larger than the depth of field, the observer can focus up and down to observe all the particle and obtain an impression of its overall form, even though he cannot observe all of it at any one microscope setting. Furthermore, because the eye is a variable focusing device, it can produce sharp images of the object on the retina for object planes that would otherwise be somewhat out of focus.

Because of the conflict between increasing resolution and decreasing depth of field, it is usually advantageous to use an objective with as low a numerical aperture as will provide the needed resolution. This is especially true since objectives of high numerical aperture are usually designed to provide a large magnification to take advantage of the high resolving power, and this high magnification results in a very small field of view. Matching the numerical aperture to the size of object is especially important, and especially easy when the object consists of particles, and if there is a wide size distribution several objectives of differing numerical aperture can be used.

Resolving powers calculated from numerical apertures indicate only the resolution theoretically attainable. Aberrations in the objective can prevent the attainment of this limit. At present, glass can be accurately ground only to produce surfaces that are planes or portions of spheres. A requirement for objectives is that all the rays of light be brought

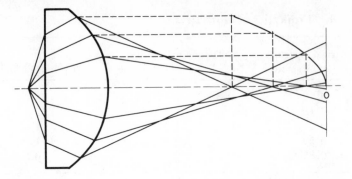

Figure 3-6. Spherical aberration at three different levels of intersection in monochromatic light. O is the point of intersection of a paraxial ray (ray close to the axis) with the optical axis. If the intercept lengths associated with several levels of intersection are determined and entered on graph paper with the vertical axis passing through O, a number of points are obtained on a paraboloid curve, which reveals immediately the magnitude of the spherical aberration for any level of intersection. Courtesy of Ernst Leitz, Inc.

together at a single focal point. However, as shown in Figure 3-6, a single spherical lens will not do this, and such a lens will not reveal fine structure. This failing is known as spherical aberration. It is not particularly noticeable for lenses of small angular aperture, but it becomes increasingly objectionable as the angular aperture is increased and therefore must be corrected to obtain high resolution.

In principle, correction for spherical aberration could be achieved by grinding a lens so that the radius would increase with distance from the center and thus decrease the diffraction of the outer rays. Such lenses can be ground for some purposes, for example, for condensers for illumination, but not with the accuracy required for objective lenses (a tolerance of less than $\lambda/4$). Instead, spherical aberration is corrected by combining two or more lenses (Figure 3-7).

The correction must take account of cover-glass thickness, and high numerical aperture dry objectives are quite sensitive to this thickness. Most objectves are designed to be used with cover glasses 0.18-mm thick, and departures from this value will result in loss of contrast. Cover-glass thickness is not so important to the use of immersion objectives, because the immersion oil has nearly the same refractive index as that of the cover glass and there is a nearly homogeneous light path between the object and the front surface of the objective.

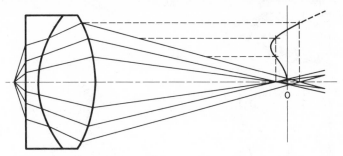

Figure 3-7. Correction of spherical aberration by means of a lens combination consisting of a collecting and a dispersing lens. The shape of the curve is radically altered compared with that on Fig. 3-6. The differences in intercept lengths have become smaller at the same levels of intersection. Courtesy of Ernst Leitz, Inc.

Chromatic aberration is very similar to spherical aberration and results from the different degrees to which light of different wavelengths is diffracted by a lens, thus bringing it to focus at different distances from the lens (Figure 3-8). Both chromatic and spherical aberrations can be decreased somewhat by decreasing the lens aperture, but this, of course, also decreases the effective numerical aperture. Objectives are corrected for chromatic aberrations by preparing the component lenses from mate-

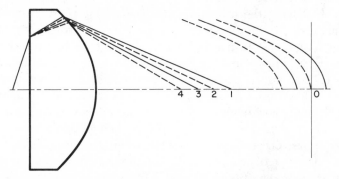

Figure 3-8. Diagram of the chromatic aberration for four different spectral colors. Because of dispersion the white light ray is fanned out into rays of various wavelengths. Accordingly a separate curve is obtained for each spectral color. The method of plotting the curves was similar to that used in Fig. 3-6. Line 1 = red light; dashed line 2 = green light; line 3 = blue light; dashed line 4 = violet light. Courtesy of Ernst Leitz, Inc.

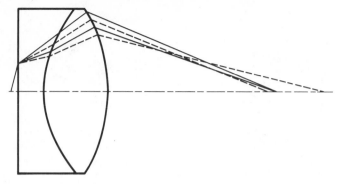

Figure 3-9. Correction of chromatic abberration by means of a lens combination. Courtesy of Ernst Leitz, Inc.

rials having different specific dispersions and refractive indexes (Figure 3-9). The combined lenses constituting the objective focus the light of two wavelengths to the same point. The images for the other wavelengths will be nearly focused to this point. There may still be faint-colored halos around the images when the illumination is by white light, but for most purposes this residual color is not objectionable. The deficiencies become especially apparent when high-magnification oculars (eyepieces) are used. These lens systems are the achromatic objectives mentioned earlier (see Figure 3-10).

Objectives having the least amount of chromatic aberrations are the apochromats, the apochromatic objectives. These are corrected for three wavelengths and are constructed of lenses of fluorite (calcium fluoride) combined with lenses of optical glass. Objectives of intermediate correction are called semi-apochromatic, or fluorite objectives when fluorite has been used. Achromats are corrected for spherical aberration for one wavelength and apochromats for two. As would be expected, the apochromats are the most expensive, and the semiapochromats are intermediate in price. Because chromatic aberrations can never be completely corrected, nearly monochromatic illumination is desirable for critical work unless color rendition is desired.

Astigmatism is the objective defect consisting of a point producing an image consisting of two lines intersecting at 90°. It can be eliminated except near the margin of the field.

Lateral color is produced when the magnification varies with the wavelength. Off-axis points are then spread out into small spectra. Oculars can be used to compensate for this defect.

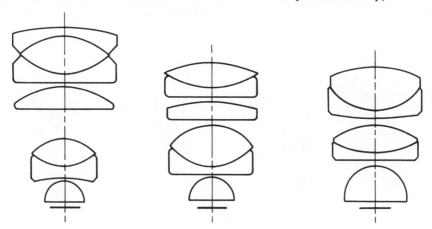

Figure 3-10.   Objective types. The objective at the right is a 4 mm, 0.65 N.A. archromat having 5 glass elements, and no fluorite. The central objective is a 4 mm, 0.85 N.A. semi-apochromat, having 5 glass lenses and 1 fluorite lens. The objective at the left is a 4 mm, 0.95 N.A. apochromat, having 5 glass elements and 2 fluorite lenses. From Encyclopedia of Microscopy by Clark. Copyright 1961 Litton Educational Publishing Co., Inc. Reprinted by permission of Van Nostrand Reinhold Company.

Coma is the production of a comet-shaped image of a point and results from portions of the lens system at different distances from the axis producing different magnifications. In a well-made microscope it will only appear, if at all, near the margin of the field.

Curvature of field consists of a well-focused center of the field accompanied by an out-of-focus remainder, and vice versa. This is especially objectionable for photomicrographic work. Special flat-field objectives are sometimes used, and some compensation can be made by the proper choice of eyepiece, especially the use of negative lens systems. This author's experience is that curvature of field is often the most obvious, irritating, and difficult to correct of the aberrations. One that is seldom found in good microscopes is distortion, which consists of the imaging of straight lines as curved ones.

1.1.3.   Oculars

The eyepiece (or ocular) plays several roles in addition to producing on the retina an image of the real image formed by the objective (Figure 3-11). It can be used to form an image on a photographic emulsion for photomicrography or on a screen for microprojection. It magnifies the

Figure 3-11. Two pairs of matched eyepieces. Courtesy of the American Optical Co.

real image formed by the objective, and superimposes images of scales, crosshairs, and the like in the ocular on the image it produces of the object.

Two common types of eyepieces are shown in Figure 3-12. The negative (Huyghenian) eyepiece is the most popular for general use. It is formed from two plano-convex lenses, one at each end of the eyepiece, with a field diaphragm between them. These simple eyepieces are quite inexpensive and work well with low and medium numerical aperture objectives.

The positive (Ramsden) ocular uses a single lens combination. Two plano-convex lenses may be used, both above the diaphragm or, as shown in Figure 3-12B, an achromatic doublet or triplet may be used. Figure 3-12B represents a "compensating eyepiece" which is designed to compensate for the residual aberrations of apochromatic objectives. The

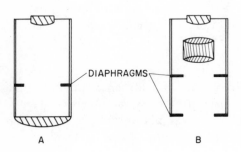

DIAPHRAGMS

A   B

Figure 3-12. Two common types of eyepieces. A is the Huyghenian (negative) and B is the compensating Ramsden (positive) eyepiece.

chromatic aberration of such eyepieces is designed to compensate for that of the objectives; thus they should be selected on the recommendation of the manufacturer of the objectives. When such eyepieces are used with achromats they produce an objectionable coloration.

Special flat-field eyepieces are available that compensate for the curvature of field produced by the objective and are designed for use with achromats. Negative flat-field lens systems have been designed for photomicrography which, unlike the other flat-field eyepieces, cannot be used for visual work. These are intended primarily for use with apochromats, and their color correction is similar to that of compensating eyepieces. They use concave lenses and produce a magnification about that of a 10× ocular.

Eyepieces that have crosshairs and eyepiece micrometers placed on the eyepiece diaphragm should be designed to permit focusing the scale and the image of the object together so that the image and the scale can be viewed or photographed simultaneously. This focusing is achieved by moving the upper lens (eye lens) up or down.

A number of types of oculars are used for special purposes. One of the most widely used has the axis inclined to the vertical for somewhat easier viewing. Of course, the entire microscope, except the stand, can often be tilted for the same purpose. Another common eyepiece is often used in connection with photomicrography. The light from the objective is partially or completely (during viewing) diverted from the axis of the microscope into a second ocular for viewing the object while the camera is in place. Such eyepieces are especially useful for locating a field to photograph, and generally one can use the microscope without having to remove the camera. They are designed with the intention that when the object appears focused in the eyepiece, the real image will be focused on the photographic emulsion. However, because of the accommodation of the eye and variations in eyes among different persons, not everyone can successfully use the auxiliary ocular for photographic focusing. Focusing on ground glass, as discussed later, is more reliable.

Binocular eyepieces are accessories that can be fitted to almost any monocular instrument, and they help to relieve eye strain, especially during long periods of observation (Figure 3-13). Other devices are demonstration eyepieces that permit two persons to simultaneously view an object with a microscope and comparison eyepieces that permit a view of two objects through two microscopes at the same time. The two microscopes, which must have similar lens systems, are placed side by side, and the eyepiece is mounted over both. The light rays from the two microscopes are brought together with two prisms so that one-half of

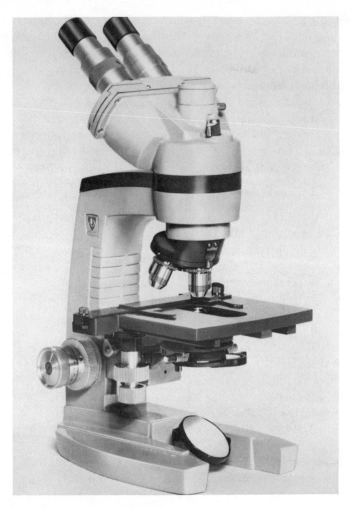

Figure 3-13.   Microscope equipped with an inclined binocular eyepiece. Courtesy of the American Optical Co.

the field of view is from one instrument and the remaining one-half is from the other.

1.1.4.   Condensers for Transmitted Illumination

The simplest illuminating system is a flat or concave mirror below the object that reflects the light from some source onto the object. The result is transmitted illumination, but such a simple system is entirely inadequate for achieving the resolution that fine objectives can produce and also may not provide sufficiently intense illumination. A much better method for obtaining bright-field illumination is with the use of a substage condenser. This is a lens system mounted beneath the stage that concentrates a cone of light onto the object. It is normally provided with an iris diaphragm to control the angle of the cone of the illuminating light. With the diaphragm closed down considerably, contrast is often improved, at the expense of resolution, as explained above. Thus the observer can achieve a compromise between resolution and contrast for each object he examines.

A two-lens Abbe condenser having a numerical aperture of about 1.2 is the most common (Figure 3-14). A four-lens Abbe condenser having a numerical aperture of about 1.4 is also used. Condensers can be considered to be inverted objectives, with light passing through them in an inverse direction and concentrated on the object.[2] To achieve the maximum numerical aperture, the space between the top of the condenser and the bottom of the microscope slide should be filled with immersion oil. Most Abbe condensers are built so that the upper lens can be removed by unscrewing it, or better and more common, so that it can be swung out of the way by the touch of a finger on a small lever. The single lens, with its longer focal length, will produce a larger image of the light source, but the light rays will be less convergent.

Some Abbe condensers have a variable focus, the position of the top lens being fixed relative to the stage and the lower lens movable in a vertical direction. This arrangement permits the observer to produce very bright, highly convergent illumination when using high numerical aperture objectives, and to produce a larger illuminated area when working with low-power objectives by lowering the bottom lens of the condenser.

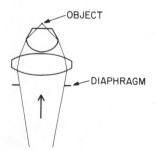

Figure 3-14. Two lens Abbe condenser.

The Abbe condensers are usually uncorrected for spherical or chromatic aberrations. However, more complex, achromatic condensers are available that are corrected for these aberrations. Achromatic condensers are especially useful for photomicrography when using high numerical aperture objectives, since they give more color-free illumination.

Dark-field transmitted illumination is produced by allowing only that light to enter the objective that is scattered by the object. Thus the illuminating rays must all be at such an angle to the axis of the microscope that they cannot enter the microscope objective. The background appears black, and the objects appear to be self-luminous. A simple way to achieve such illumination is to place an opaque disk (stop) in the center of the diaphragm opening beneath the condenser. This will only be effective, of course, if the rays from the outer portion of the objective are sufficiently oblique. Thus a simple Abbe condenser can be used with 8 mm and 16 mm objectives, but a higher numerical aperture condenser must be used for higher-powered objectives.

Special condensers are available for producing dark-field illumination. The most common of these are the parabaloid condensers, which produce a hollow cone of light with numerical apertures between about 1.15 and 1.40. These condensers (or illuminators) are essentially a piece of glass having the form of a portion of a parabaloid of revolution and equipped with a central stop. The curved surface is silvered; parallel rays entering the condenser are reflected and converge to the focal point (Figure 3-15).

Parabaloid surfaces cannot be ground with great accuracy, and a somewhat better device is the cardioid or bispheric condenser which employs two spherical reflecting surfaces. A cylinder of entering rays is reflected

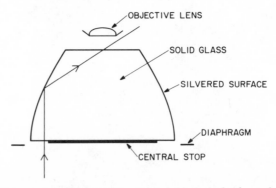

Figure 3-15.   Schematic diagram of a parabaloid dark field condenser.

from the first surface (a spherical cavity, concave upward) onto a second silvered surface, which is the curved outer surface of the condenser, concave inward. The second reflection delivers a cone of light to the field of view of the microscope. Such condensers have numerical apertures of 1.20 to 1.40. Both types of dark-field condensers are designed to be used with immersion oil between the condenser and the microscope slide.

Dark-field illumination causes particles smaller than the resolving power of the microscope to appear as unresolved dots of light. This can be a disadvantage because minute extraneous particles, and scratches and other imperfections on the microscope slide are annoyingly apparent, and even out-of-focus extraneous material can appear as blurs of light. On the other hand, for some purposes counting the total number of particles, resolved and unresolved, is desirable; then dark-field illumination is especially appropriate. This is a form of ultramicroscopy, the "ultra" referring to the visibility of the unresolved particles.

In its original form, the dark-field illumination of ultramicroscopy was usually provided by a beam of light at 90° to the axis of the microscope and directed into a glass cell beneath the objective. This cell contained the particles to be counted in suspension in some fluid. This was the arrangement of the slit ultramicroscope of Siedentopf and Zsigmondy which was used for much of the early research on colloidal particles. A beam of light was focused on a horizontal slit and an image of the slit was focused at the field of view of the microscope, in the cell containing the colloid.

Whytlaw-Gray and his co-workers[9] were probably the first to use the slit ultramicroscope to study particles suspended in gases without removing them from suspension. The viewing microscope had to be placed horizontally to prevent the particles from settling out of focus. For this purpose cardioid or parabaloid dark field condensers were useless. Also, the cell could not be too small, and it had to be possible to change the contents rapidly. The measurements (counts) were made with a continuous flow of the aerosol up through the cell. Errors were found in this procedure; for instance it was not possible to distinguish between faintly illuminated particles outside the light beam and very small particles within it. Therefore, a new cell was designed such that the depth of the layer counted was determined by the separation of the parallel surfaces of the cell. Obviously, long focal length objectives were necessary, and Whytlaw-Gray used a 25 mm, 0.3 N.A. objectve. Whytlaw-Gray and his co-workers used the ultramicroscope mainly to study coagulation of the particles in aerosols, and a source of error not connected with microscopy was the loss of particles by diffusion to the walls of the cells.

The slit ultramicroscope, with its orthogonal illumination, is seldom used today, and the annular illumination produced by parabaloid or cardioid condensers has essentially replaced it. Special cardioid condensers and cells are available for ultramicroscopic work when the use of a suspension is appropriate. The cells are very thin and, as Whytlaw-Gray observed, are not appropriate for suspensions in gases.

There are a number of ways in which ultramicroscopes can be used for investigating aerosols in addition to studying the rates of coagulation. If the volume from which the particles were collected on a microscope slide can be estimated, the count will indicate the number concentration in the aerosol. If the mass concentration of particles is also known, the average mass per particle can be calculated and from this the size of the particle of arithmetic mean mass. The degree of Brownian motion of particles suspended in a gas, observed with some variation of the slit ultramicroscope, can be used to estimate the size of a particle.

Various devices, specially designed, have been used to collect and observe the light scattered by individual small airborne particles against a dark field. However, the light collecting devices often resemble small telescopes rather than microscopes. Such equipment can be used to investigate the thermal forces on particles, the charges on particles, and various other aspects of aerosol science. Except for such special instrumentation, ultramicroscopy even with annular illumination is seldom used today, probably because of the general availability of electron microscopes.

1.1.5. Incident Illumination

Most studies of fine particles by microscopy use transmitted illumination, but for some purposes incident illumination is preferable, either alone or in conjunction with transmitted light. For example, when particles are opaque, transmitted illumination reveals only an outline of each particle, whereas incident light may resolve surface details. Obviously, incident illumination must be used to examine particles collected on an opaque surface unless the particles are removed and remounted on a microscope slide, an undesirable procedure if it can be avoided.

The simplest (and least satisfactory) method for achieving incident illumination is to shine on the object the light from a lamp located beside and above the microscope stage. This approach is only satisfactory for examination with very low power, long focal length objectives, and is nearly impossible with high numerical aperture objectives because of the short focal length and the corresponding obstruction at the field of view by the objective. A simple modification that produces vertical illumination involves mounting a nearly transparent reflector between the object and objective at a 45° angle to the microscope axis. A horizontal beam of light is partially reflected onto the object without preventing light from the object reaching the objective. Such techniques are of little use for measuring the particles collected from aerosols.

Numerous incident illuminators are commercially available that combine the functions of objective and condenser. One type uses the objective lens also as the condenser, a horizontal light beam being deflected down through the objective. Just as the semi-reflecting plate can be placed below the objective, it can be placed within the objective mounting as shown in Figure 3-16 (Beck illuminator). The entire aperture of the objective is used and bright-field illumination is obtained. A semi-transparent metal coating is often used to provide the reflecting surface, and additional coatings of transparent substances may be used to avoid multiple reflections. Some commercially available illuminators of this sort are quite elaborate and are equipped with side tubes containing the light source. This tube in one such commercial instrument includes, in addition, to the lamps, (a) a diffusing screen to render the light homogeneous, (b) a lever to set the aperture stop, (c) a lever to set the radiant field aperture, (d) pins for centering the radiant field stop, (e) a filter cell to hold rings into which light filters or an azimuth diaphragm can be placed (the polarizer for vertical illumination can also be placed here), (f) a lever for vertical setting of an aperture stop, (g) an auxiliary mount for greater stability, (h) an auxiliary lens and filter slide, (i) a swing-out neutral glass filter for rapid variation of the illumination intensity, and (j) a bright-field/dark-field change slide for a bright-field and dark-field annular diaphragm. Its use for dark-field illumination is described below.

Illumination is sometimes obtained with a total reflector or prism, as shown in Figure 3-17. Such illuminators reflect the entering light very efficiently but obstruct some of the image-forming rays and thus somewhat decrease the numerical aperture.

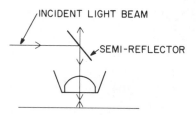

Figure 3-16. Incident light illuminator using a semi-reflecting plate.

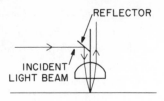

Figure 3-17. Oblique incident light illuminator.

Dark-field illumination is also produced with a combined condenser-objective in which the condenser is an annular diffractor or reflector around the objective lens. Examples are the Zeiss Epi-Condenser and the Leitz Ultropak. In this way very bright, very oblique illumination can be obtained, and only light scattered by the particles can enter the objectives. The Ultropak depends entirely on a lens system, whereas in the Epi-Condenser the light surrounding the objective is reflected onto the object with a polished steel surface (Figure 3-18).

1.1.6.   Critical and Köhler Illumination

It is almost a truism that proper illumination is required to make full use of the capabilities of a microscope. Yet most microscope users do not bother to employ, or are not familiar with, methods for producing illumination consistent with the sophistication of the instruments. Two meth-

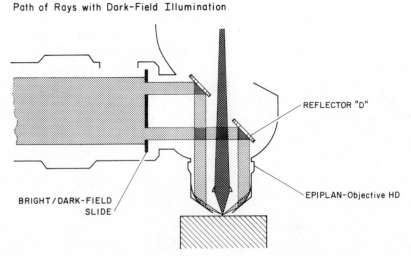

Figure 3-18.   Path rays in the Zeiss Epi-Condenser. Courtesy of Karl Zeiss, Inc., New York.

ods for achieving fine illumination have had widespread use: critical illumination and Köhler illumination. The latter is generally used at present, but both are described below.

For critical illumination, an image of the light source is produced in the plane of the object examined. This can be accomplished entirely with the substage condenser, but only a small portion of the light from the lens will illuminate the condenser. The field will then be too dim under many conditions, for example, when using dark-field illumination or high magnifications or when taking photomicrographs. Therefore, most microscope lamps provide an auxiliary condenser, in front of the light source, that can produce a parallel beam. This beam is reflected up into the condenser with a flat substage mirror; the beam should be well centered on the mirror. If the parallel beam of light is wider than the diaphragm opening of the substage condenser, it can be made slightly convergent with the auxiliary condenser, which is movable in most microscope lamps, but the beam should completely fill the aperture of the substage condenser. With the use of a low-power objective, the microscope is focused on the particles, and the substage condenser is raised or lowered until a sharply focused image of the light source is seen through the microscope. The angular aperture of the condenser is then made equal to that of the objective used for observing the particles. This is accomplished by focusing that objective on the particles and removing the eyepiece. Looking down the tube one sees the illuminated back of the objective. The condenser diaphragm is then slowly closed until a contrasting circle of light is seen at the back of the objective. The diaphragm is then opened until the back lens of the objective is just filled with light. The angular aperture of the illuminating beam will then be the same as that of the objective. Critical illumination will then have been achieved, but this procedure must be repeated for each objective.

When using a high numerical aperture substage condenser with a low numerical aperture objective, critical illumination may not fill the field. This difficulty can often be remedied by removing the top lens of the condenser (especially easy when the condenser is equipped with a top lens of the swing-out type) and repeating the procedure to obtain critical illumination. Removing the top lens of the condenser decreases the numerical aperture of the condenser, as does closing the diaphragm of the condenser. Of course, the area illuminated can also be increased by increasing the area of the source of light. Ordinary incandescent lamps are not satisfactory for critical illumination because the filaments are of such a small diameter that their image will not fill the field. Special lamps have been designed for critical and Köhler illumination, such as those using ribbon filaments. A diffuse source such as a frosted bulb is not very satisfactory because it does not produce a parallel beam. Thus some of the light is not sharply focused on the particles, and the resulting glare causes a loss of contrast and resolution. Critical illumination will nearly always completely illuminate the field when using high numerical aperture objectives.

Köhler illumination differs from critical illumination in that (a) an image of the light source is formed in the plane of the substage condenser rather than in the plane of the object observed, and (b) an image of a field diaphragm in front of the light source is produced in the plane of the object (Figure 3-1). The image of the light source, produced with the auxiliary condenser, should be just large enough to uniformly fill the aperture of the substage condenser. Therefore, this image should usually be somewhat larger than the light source (filament). If the image is too small, the effective numerical aperture of the substage condenser will be decreased. The focusing can be achieved by closing the iris diaphragm of the substage condenser and using a mirror to observe the image on the diaphragm while moving the auxiliary condenser with respect to the light source. The microscope mirror is then adjusted to center the light source image on the substage diaphragm. If the light source is built into the stand, the centering can be accomplished with the centering screws on the light source. Then the substage diaphragm is opened.

Next, if one uses a low-power objective and looks through the microscope, the image of the nearly closed diaphragm of the lamp is focused on the field of view with the substage condenser. With the objective to be used in place, the lamp diaphragm is opened until the field of view is just filled with light. The ocular is then removed and the diaphragm of the substage condenser is adjusted until the back lens of the objective is just filled with light, as in critical illumination. Some microscopists, including the author, prefer to fill only about three-fourths of the back lens of the objective with light. Although this procedure somewhat decreases resolution it can greatly increase the contrast.

Sometimes the intensity of light produced by critical or Köhler illumination is too great. It should not be reduced with the diaphragms or by lowering the substage condenser. Instead it should be controlled by neutral density filters or by decreasing the voltage across the lamps. The latter procedure should not be used for color photomicrography because it will change the color temperature of the light (discussed in Section 1.8).

A major advantage of Köhler illumination is that the field of view of the microscope is evenly illuminated because the light is a reduced image of the lamp (field) diaphragm. Therefore, unlike critical illumination, a wide light source is not required. Other advantages are that glare is reduced to a minimum, and the illumination can be used efficiently for objectives of various powers since the opening of the field diaphragm can be reduced in diameter as the numerical aperture of the objective is increased.

Both critical and Köhler illumination can be achieved with vertical illuminators. Critical illumination, as discussed above, involves producing an image of the light source at the object plane using the substage condenser. But with vertical illuminators that use the objective as the condenser, the position of the objective is determined by focusing on the object and cannot be changed to form the light source image. Chamot and Mason[5] describe three ways to deal with this restriction:

1. The light source is placed at the same "optical distance" from the objective as the image distance.

2. An image of the light source is produced at the same distance using an auxiliary condenser. The objective will then reproduce this image in the object field.

3. An additional lens in front of the light source, together with the objective, produces an image of the light source in the object field.

Aperture and field diaphragms are required for vertical illuminators if critical illumination is to be used. These are provided in the better commercially available vertical illuminators. For uniform illumination, a ribbon-filament light bulb is preferable; a coiled wire filament produces very uneven illumination.

Köhler illumination is especially appropriate for vertical illuminators because it avoids the difficulties mentioned above. Since the objective serves as the condenser, an image of the light source is projected on the back of the objective and must fill its aperture. This is accomplished with an auxiliary lens in the side tube. A field diaphragm placed at a distance from the objective about equal to the image distance is imaged by the objective in the field of view. An aperture diaphragm in the side tube can be used to control the convergence of the rays and achieve the compromise between resolution and contrast described above.

Some vertical illuminators are not equipped with the auxiliary condensers and the diaphragms required for producing Köhler or critical illumination. Even illumination, at least, can be achieved by adjusting the focus of the lamp, or by placing a ground glass diffusing plate in front of the lamp.

1.2. Phase Contrast Microscopy

A major problem in microscopy when examining objects of inherently low contrast is achieving satisfactory contrast without at the same time producing an unacceptable decrease in resolution. A very popular method for increasing contrast without decreasing resolution is the use of phase-contrast microscopy. It has proved to be very useful for the examination and photomicrography of biological specimens, but this author has not found it very satisfactory for the measurement of particles. The method tends to produce a rather annoying halo around the image of each particle that somewhat obscures the particle boundary. The method can be useful for particles that would be difficult to view by ordinary microscopy because of low contrast, but otherwise it should probably not be used for fine particle measurement. Bennett et al.[10] state that particle counts with phase microscopy are more accurate because of the increased visibility of the particles and are more complete because smaller particles can be seen. They also suggest that with the proper choice of mounting medium, differential particle counting can be achieved for components of a mixture that look alike with more conventional illumination.

Consider a small object at the center of the field of view of a microscope and illuminated with axial transmitted illumination. Some of the light passing the particle will be unchanged in direction, passing through the center of the microscope, but light will also be diffracted at the edge of the object and take a different, oblique path through the objective. This separation of paths makes possible the modification of the image-forming rays, the background-forming rays, or both from each point to increase contrast. In phase-contrast microscopy both the phase and amplitude relationships are changed, and the central and oblique rays are brought together at the real image where reinforcement or interference occurs as shown in Figure 3-19. The result is a variation in brightness for regions of differing refractive index. The halos mentioned before indicate the boundaries of regions of differing refractive index and thus the physical boundaries in the object observed.

As described earlier, axial illumination produces less resolution than oblique illumination, but obviously highly oblique illumination cannot be used for phase contrast because then separation of the direct from the

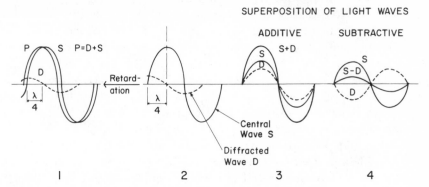

Figure 3-19.   Contrast control by adding or subtracting light.[11]

diffracted rays would not be possible. Therefore, a compromise is made, commonly by using an annular, and thus hollow, cone of illumination produced by a diaphragm mounted below the condenser. This diaphragm consists of a central stop, an annular transparent ring, and an annular stop beyond the transparent ring, as shown in Figure 3-20. Since the condenser focuses this light on the object examined, the undiffracted light becomes a diverging hollow cone between the object and the objective. The region intercepted by this cone at the back aperture of the objective is called the conjugate annulus. The remainder of the back aperture of the objective is called the complementary area, and it is through this region, both inside and outside of the conjugate annulus, that light diffracted by the object passes. A phase plate, also called a diffraction plate is placed just behind the objective as shown in Figure 3-20 and serves to alter the phase and the intensity of the rays passing

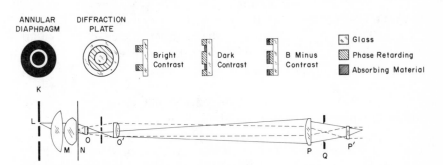

Figure 3-20.   Diagram of the Spencer phase microscope.[11]

through it. Note that it is divided into regions of differing properties corresponding to the conjugate annulus and the complementary area. Various types and degrees of contrast can be produced by using various types of screens. If the central rays are retarded by $\lambda/4$, bright or negative contrast is produced in which the more highly refractive portions of the object appear lighter (Figure 3-19). Conversely, if the diffracted rays are retarded, dark or positive contrast is produced. The contrast can also be increased by differentially decreasing the intensities of the rays with absorbing films on appropriate parts of the phase plate.

A considerable assortment of equipment is available for phase-contrast microscopy. Achromatic objectives having magnifications of $10\times$ through $100\times$ are manufactured, the highest magnifications being for oil-immersion objectives. Each of these has a built-in phase plate back of the objective lens. Fluorite and apochromatic objectives are also available for the best work.

A special objective is the Wilska phase-contrast objective which uses finely divided carbon particles to decrease the intensity of the central rays. This approach nearly removes the objectionable halo surrounding the image of each particle. Needham[12] recommends bright (negative) contrast for the examination of fine particles.

The substage condenser for phase contrast work is equipped with an off-center rotatable disk carrying several annular stops (diaphragms), any one of which can be turned into position beneath the condenser lens in its focal plane. Each of the stops corresponds to a particular objective because the condenser annulus and the ring of the phase plate must be matched. A completely transparent opening is also provided in the disk beneath the condenser for adjusting the illuminating system, since Köhler illumination should be used. Some manufacturers provide a single stop and "change" its size with a zoom lens in the condenser.

An accessory is a small telescope that can be placed at the top of the drawtube in place of the ocular and is used to observe the back of the objective lens while superimposing the image of the condenser annulus with the ring of the phase plate. A green filter provides green illumination, because the phase differences are calculated for this wavelength range.

The first step when using phase-contrast microscopy is to produce Köhler illumination as described above. The desired objective and the corresponding annular diaphragm are put in place, and the microscope is focused on the object. The small telescope is then substituted for the eyepiece and is focused with the telescope eye lens until the annulus of the condenser diaphragm and the ring of the phase plate are sharply imaged. These two images are made to coincide, usually with centering screws on the substage condenser assembly. The ocular is then replaced.

Most conventional microscopes can readily be used for achieving phase contrast. The better polarizing microscopes are usually equipped with a Bertrand lens which can be slid into the body tube for observing interference figures produced at the back of the objective by anisotropic crystals. This lens can be used instead of the small telescope. The OPTOVAR magnification changer on Zeiss microscopes which normally is used to increase the magnifying power of the oculars can also be used to center the annulus.

Phase-contrast incident illumination can also be produced. Because the objective serves as the condenser, the annular diaphragm is placed in the side tube where it will not interfere with the rays diffracted by the object. A single annular diaphragm can be used and an auxiliary lens adjusted to provide the proper dimensions of the image of the annulus for each objective. The phase plate is located either just back of the objective or at the image of the back of the objective aperture produced by a projection system in the body tube.

1.3. Interference Microscopy

Interference microscopy, like phase contrast microscopy, increases contrast by producing differing degrees of interference depending on the optical thickness of various parts of the object observed. For particle measurement it has the advantage over phase-contrast microscopy of not producing halos around the images. According to Francon,[13] who has written an excellent review of the theory and types of equipment for interference microscopy, phase-contrast microscopy is effective only when the detail observed is small because otherwise the diffracted light is to a considerable extent coalesced with the direct light. Interference microscopy is not subject to this limitation.

The principle of the interference microscope is very simple, and is that of the interferometer (Figure 3-21). A ray of light originating from one point of a light source is split into two spatially separated parts. One part is retarded by a transparent object which may be the object observed with the microscope or a plate inserted for the purpose. The rays are then recombined, and the phase shift usually causes interference and thus a decrease in light intensity. If the difference between the paths is an odd number of times $\lambda/2$, complete extinction occurs. A dark field can be produced with such a phase shift by placing an appropriate

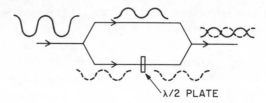

Figure 3-21.   Mechanism of interference production in the interference microscope.

retarding plate in one path. Now if the object observed is placed in the
other path, the difference between the paths will no longer (usually) be
an odd multiple of λ/2, and the image of the object will appear bright
against a dark background. Of course, the field does not have to be dark
to obtain contrast.

Numerous methods have been used to split the rays. Perhaps the
simplest, which is used in Dyson's microscope, is the semi-reflecting
(half-silvered) mirror which reflects part of the light beam while allowing the rest to pass. The Leitz interference microscope uses prisms and
two identical objectives. Polarization interference microscopes split the
incident ray by means of a birefringent system.

Incident illumination interference contrast can also be produced, and
the author has found it especially useful for observing the surface
detail of large particles, which can help identify them (Figure 3-22).
Commercially available incident illuminators have accessory equipment
for this purpose.

1.4. Ultraviolet Microscopy

The ultraviolet microscope was originally designed to provide considerably greater resolving power than that which can be obtained with
microscopes designed for visible light. It must include some means for
rendering visible the image produced in ultraviolet light. Optical glass
transmits near ultraviolet light, and some improvement in resolution
over ordinary microscopes can be obtained with such illumination, but
much greater resolution can be obtained using even shorter wavelengths
and an optical system that can transmit them. For such short wavelengths, lenses can be ground from fused quartz or fluorite, and curved
mirrors can replace some or all of the conventional optics.

Today, the ultraviolet microscope is little used for its high resolving
power because of the very much higher resolving power of electron

Figure 3-22. Photomicrograph taken with interference incident illumination of particles collected from Kilauea volcano eruption fume. The particles have crystalized from coalesced droplets collected by impaction on a platinum slide. Distance across the micrograph is 130 μm. Reprinted with permission from Cadle et al., *Geochimica et Cosmochimica Acta* **35**, 503 (1971), Pergamon Press. 700×

microscopes. However, it is used occasionally for studying biological material because of the different ultraviolet transmissivities of different types of such material. For example, various normal and abnormal cells can be differentiated with ultraviolet microscopy. Although ultraviolet microscopes are seldom used for particle counting or measurement, if one is available it can be very useful because of its high resolving power.

Perhaps the most popular objectives for ultraviolet microscopy have been those designed by Grey and Lee.[14,15] For example, one of these is corrected from 2200 Å to the near infrared and has a numerical aperture of 0.72 (Figure 3-23). It uses both a mirror and several lenses of fluorite and fused quartz. The large wavelength range for which the objective is corrected permits focusing in the visible for photography in the ultraviolet region.

The early ultraviolet light sources were cadmium spark lamps, which were both noisy and odorous.[8] However, these have been replaced by mercury arc lamps. For biological work, when it is desired to examine the specimens with different wavelengths, grating monochromers are used. Otherwise, a filter that transmist the 3560 Å line is desirable for the objectives that have been designed to focus for this wavelength.

Until the development of objectives that were corrected for the visible as well as the ultraviolet regions, and also the development of television image scanning, focusing was largely a matter of trial and error, recording the images as photomicrographs. Fluorescent screens were also used to render the images visible but the results were grainy and dim. Today it is possible to focus with visible light, change to ultraviolet light, and take photomicrographs on polaroid film, obtaining a permanent record in a matter of seconds. The auxiliary lenses such as condensers and those used to focus the image on the photographic film must transmit ultraviolet radiation.

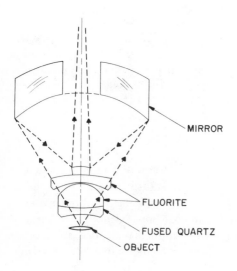

MIRROR

FLUORITE

FUSED QUARTZ

OBJECT

Figure 3-23. Lens components have been added to the Schwarzschild mirrors to provide an objective of numerical aperture 0.72, corrected for use in the wavelength region from 220 mμ through the visible spectrum. Nine percent of the objective aperture area is obscured by the convex mirror.[15]

The television scanning techniques for displaying the images are very satisfactory but rather expensive. One is based on the flying-spot principle which has been used with visible light microscopy to count and measure particles. It makes use of the microscope acting in reverse, the light source being an ultraviolet-emitting cathode ray tube, the face of which is located in the plane where photographic film would be placed for photomicrography. The rays from this tube are focused on the object to be studied, and the rays transmitted by the object impinge on an ultraviolet-sensitive photomultiplier tube. The cathode ray tube is not completely illuminated at any given moment but rather produces a flying spot of light, as in conventional television, which probes the entire field of view of the microscope in a very short time. The electrical signal from the photomultiplier tube is converted to an image of the field of view of the microscope on a cathode ray tube as in conventional television. As discussed later, the signal from the photomultiplier can be electronically processed in various ways to provide information concerning the numbers of particles in the field of view and concerning the particle size distribution. The chief disadvantage to this technique, in addition to its expense, is the difficulty of varying the wavelength of the ultraviolet light from the cathode ray tube. However, this is not a serious drawback for fine particle work.

A more conventional use of television techniques is to use the microscope in the usual manner, producing the ultraviolet image on an ultraviolet-senstive image-orthicon, and transmitting the image by closed circuit television to a conventional television screen. Again, the signal can be electronically processed to provide particle number and size distribution information.

Another method for seeing the ultraviolet image makes use of an ultraviolet image-converter tube. A photoemissive cathode produces the image in electrons on a fluorescent screen, which in turn produces the visible image. The RCA Corporation image-converter tube is called the Ultrascope.

1.5. Microscope Design

1.5.1. Illumination

Until recent years, illumination for the microscope was provided by a separate lamp, the light for transmitted illumination being reflected into the microscope with a mirror mounted beneath the condenser. Some such microscopes are still manufactured, and the cheapest ones are not

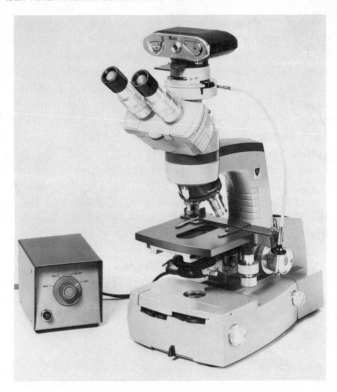

Figure 3-24.   Microscope with a light source built into the base and equipped for photomicrography. Courtesy of the American Optical Co.

even equipped with substage condensers. Except for the crudest work with particles, the separate lamp should be equipped with a focusable lens and a field diaphragm to permit the use of Köhler illumination. As mentioned earlier, if critical illumination is to be used, a ribbon filament is preferable to a wire filament to provide uniform illumination.

To an increasing extent, the horseshoe base and mirror arrangement used for so many years is being replaced by the manufacturers with a base containing a built-in light source. The source is permanently aligned, always immediately available, and designed for Köhler illumination (Figure 3-24). The low-voltage lamp uses a coiled filament and in the better systems can produce a brilliant light for photomicrography, dark-field work, phase contrast, and so forth; voltage taps on the transformer and filters can be used to obtain various intensity levels.

Another arrangement is shown in Figure 3-25, which shows the light path through the Zeiss Universal microscope. Both transmitted and incident illumination systems are built-in.

1.5.2.   The Arm

The older microscope bodies are designed to tilt back about a pivot where the arm meets the base for viewing convenience. However, the newer instruments with built-in light sources are not designed to tilt, but instead are designed to accept monocular or binocular inclined eyepieces. Unlike the older microscopes with their conventional microscope tubes, the upper end of the arm may be equipped with an inclined plate and with a circular hole in the center, designed to accept a variety of interchangeable microscope components. On some microscopes a horizontal ring rather than an inclined plate is used for the same purpose.

Regardless of the general design of the microscope, rigidity throughout is essential, and the center of gravity should be low enough to provide stability. Chamot and Mason[5] recommend against microscopes constructed from zinc-base die castings because eventually they tend to crack, swell, and disintegrate.

1.5.3.   The Body Tube

The better microscopes of the older designs were provided with a graduated extension called the draw tube, which made possible tube-length adjustments. There are a number of advantages to this arrangement. For example, some ocular micrometers are calibrated by the manufacturer for a particular objective and tube length. It is also useful when calibrating ocular micrometers with stage micrometers. If both micrometers consist of series of lines, the tube length can be adjusted so that a whole number of spaces between the lines of one micrometer corresponds to a whole number of spaces on the other.

Binocular eyepieces that are really binocular bodies in the modern microscopes have already been mentioned. Triocular bodies are also available for use on many modern microscopes. They are similar to the arrangement for photomicrography, consisting of a vertical tube for projecting the image onto the photographic film and a side tube and ocular for forming an image with the eye, except that the side tube arrangement includes an inclined binocular. The rays from the objective can be directed either with a beam divider or with a prism.

An especially annoying weakness of some microscopes designed to focus by moving the body tube with a rack-and-pinion system was a

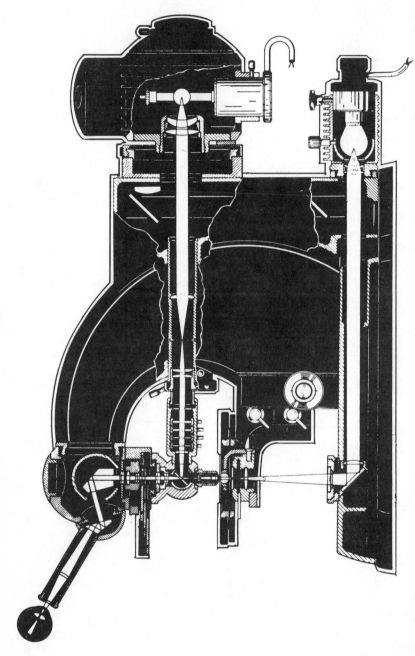

Figure 3-25. Zeiss "universal" microscope with built-in systems for both transmitted and incident illumination. Courtesy of Karl Zeiss, Inc., New York.

tendency to move downward when heavy objects such as cameras or special oculars were mounted at the top of the body tubes.

A recent completely redesigned system is shown in Figure 3-26. Its modular design permits greater versatility, and the optical axis is coincident with the axis of symmetry of the microscope for greater stability.

1.5.4. Photomicrographic Equipment

A large array of microscope modifications for photomicrography is available from various manufacturers. One type consists of a special stand on which both the microscope and camera are mounted and a special body tube with side ocular (or binocular) interchangeable with the regular body tube. Other modern microscopes are equipped with built-in cameras. The 35 mm cameras are very popular because of the large number of photomicrographs that can be produced on a single roll of film. This author prefers a larger camera with an interchangeable back that can accept a ground glass screen for focusing, cut film, or Polaroid film.

1.5.5. Objective Changers

Most microscopes are equipped with revolving nosepieces holding three or four objectives for changing quickly from one magnification (and resolving power) to another. The objectives are designed to be parfocal so that the microscope remains focused (or nearly focused) on the object in spite of changing from one objective to another. Many microscopes are now designed so that the revolving nosepiece is readily removed from the microscope body and replaced with special objectives such as those for incident illumination. Centerable fittings for objective mountings are especially important when examining crystals with polarized light, since the crystal is rotated on a rotating stage without moving it out of the field of view.

1.5.6. Focusing

Focusing is usually achieved with a rack-and-pinion coarse adjustment, and one of a number of types of fine adjustment. There should be no "play" in the coarse adjustment, and the body tube should not be permitted to move downward from the weight of accessories.

Unlike the coarse adjustment, many designs are used for the fine adjustment, including a screw and nut with worm wheel, reducing level or reducing cam, and a planetary ball-bearing drive. None of these seems to have a marked advantage over the others. On many of the

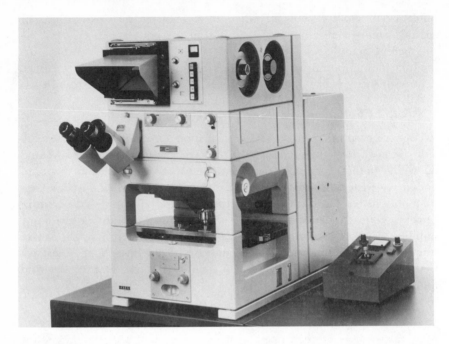

Figure 3-26. The Zeiss "axiomat." Courtesy of Karl Zeiss, Inc., New York. The Modules are from the bottom up:

Stage Module The zoom condenser slide is inserted in front. Focus and zoom control knobs are shown on the front panel. Large mechanical stage.

Objective Module with 6x nosepiece, and one of the either-side focusing knobs shown on the right side.

Observation Module with one binocular tube attached. Control knobs for beam splitter, zoom system, and centering telescope facing the operator.

Camera Module with projection screen of the 5x5″ camera in front, the casette of the 35 mm camera on the side of the module, next to the ASA setting. All push-buttons to operate both cameras in all possible modes are arranged vertically, next to the screen, topped by the meter window, indicating the exposure time.

Light Source Module mounted in back with three light sources inside. This module can be replaced by one of the illuminators of the Zeiss line, if only one light source is needed.

Control Module On the panel: On- and off-switch, voltage regulation for the light sources, remote-control of focus and remote-control of camera shutters.

176

newer microscopes the focusing adjustments have been lowered so that focusing can be undertaken with the hand resting on the table.

1.5.7. The Stage

The stage should be fairly large and designed to accommodate a mechanical stage or a hot stage. Circular, rotating stages are preferable and are essential for polarizing microscopes. The edge of the rotating stage should be graduated in degrees. A means for locking the stage against rotation and centering screws should also be provided. Some microscopes are designed so that the regular stage can be removed and replaced with a hot stage.

Focusing with many modern microscopes is accomplished by raising or lowering the stage instead of the body tube.

A mechanical stage is almost a necessity for counting and measuring particles if statistically significant results are to be obtained. Mechanical stages that are accessories rather than incorporated permanently in the stage are preferable so that they can be removed, leaving the stage free for mounting other objects such as hot stages. They should, of course, be designed to rotate with the stage, and they should have two motions at right angles to each other. The better mechanical stages are equipped with graduations and verniers so that by noting the position of a field of view it can be relocated. The graduations also make possible moving the microscope slide by equal successive amounts for particle measuring and counting.

1.5.8. Portable Microscopes

Field work involving airborne particle measurements sometimes requires particle counting and measuring without waiting to return the samples to a laboratory. Several commercially available portable microscopes have been designed for such purposes. They are usually smaller than conventional microscopes, and the base may fold back to fit into a small case. One such microscope, described in detail by Needham,[12] is the McArthur inverted microscope which measures $4 \times 2\frac{1}{2} \times 2$ in. It is quite unconventional in design, being an inverted microscope in the sense that the object is observed from beneath. In spite of its small size it includes a substage condenser and three parfocal achromatic objectives, the highest powered of which is designed for oil immersion.

1.6. Polarizing Microscopes

Polarizing microscopes are designed to identify objects such as particles by their optical properties, or to determine the optical properties of crystals of known substances. Since such applications are for the most part beyond the scope of this book, polarizing microscopes are described only briefly. Numerous excellent and extensive discussions of the applications of polarized light to microscopy are available, for example, in Chamot and Mason[5] and in Schaeffer.[16] Chamot and Mason[5] and Stoiber and Morse[17] give extensive bibliographies on the subject. A description is justified here since a polarizing microscope can often be used to distinguish among different substances in particulate form in a mixture on a microscope slide, permitting one to obtain size distributions for each substance. Furthermore, particle size distributions are often made with polarizing microscopes when the polarizing features are not used, and it is helpful to be acquainted with all the features of the instrument with which one is working.

The most important distinguishing features of a polarizing microscope, sometimes called a petrographic microscope, are the polarizer and the analyzer. The polarizer acts by rejecting the light vibrating in all but a single plane, and transmitting the light vibrating in that plane. The transmitted light is then said to be polarized. Until fairly recently the polarizer was nearly always a polarizing prism such as the nicol prism, and was constructed of the double refractive material calcite. The vibrations of light entering a crystal of calcite are resolved into two components that are polarized, vibrating at 90° with respect to each other. By properly cutting a calcite crystal into two parts and cementing the parts together with a different orientation, one component is totally reflected by the cemented surface and absorbed by a black coating. The other component passes through the prism and illuminates the object observed.

Polarizing prisms to a considerable extent have been replaced by Polaroid film. It consists of sheets of plastic material containing very large numbers of very small crystals of quinine iodosulfate. For use in polarizing microscopes the film is mounted between thin plates of optical glass. The main advantage is the considerable decrease in cost. These films act by pleochroism. Light vibrations in one direction in the filter are absorbed, whereas those perpendicular to the absorbed vibrations are transmitted.

The polarizer is mounted beneath the substage condenser in such a manner that the direction of vibration of the polarized light is always known.

The analyzer is usually mounted just above the objective, but in some microscopes it is located above the eyepiece. It is a polarizing prism or film, like the polarizer, and the direction of vibration of the light it transmits must be accurately known. Many analyzers can slide in or out of the body tube. Wherever they are mounted they should be rotatable through at least 90°, and graduations should be provided so that the direction of vibration of the light they transmit can be read from the microscope. The polarizing microscope should also have a rotating stage, a cross-hair eyepiece, and various compensators. A Bertrand lens for observing interference figures is helpful.

The polarizing microscope can readily be used to distinguish between optically isotropic and anisotropic transparent particles. The former have identical optical properties for all directions of light passage through the particles. An example is sodium chloride crystals. The latter consist of crystals that exhibit different optical properties in different directions and are said to be doubly refractive or birefringment as well as optically anisotropic. Such substances possess different indices of refraction, depending on the direction of vibration of the light passing through them. An example is calcite.

Optically anisotropic substances resolve the light entering them into components which are polarized such that they vibrate only in definite planes perpendicular to each other. The components pass through the crystals at different rates, corresponding to the different indices of refraction.

The polarizer and analyzer (together, the polarizers) of a microscope are said to be crossed when the direction of vibration of the light passing through one is perpendicular to that through the other. Then if all material in the light path between the polarizer and analyzer is optically isotropic, the field will appear dark. However, if an optically anisotropic crystal is observed between crossed polarizers and the stage on which the crystal is mounted is rotated, the crystal will appear alternately light and dark. The positions of maximum darkness (of "extinction") are separated by intervals of 90°. Except when the directions of vibration of an anisotropic crystal are aligned with those of the polarizers, the plane of vibration of light from the polarizer is resolved by the crystal into its characteristic vibration planes, and light can pass through the analyzer. When the directions of vibration of the crystal are aligned with those of the polarizers, the directions of vibration are not changed by the crystal, and the light is completely stopped by the analyzer. Because isotropic crystals produce no such effect, the polarizing microscope can readily distinguish between particles consisting of the two types of material.

Furthermore, when optically anisotropic crystals exhibit a definite crystalline form, anisotropic crystals of different substances can often be distinguished by determining the directions of extinction relative to the

crystal faces and axes. The determination is made by aligning the cross
hairs of the eyepiece with the vibration directions of the polarizers.
Then the positions of extinction relative to the cross hairs are observed.
Parallel extinction occurs when the long direction of the crystal is
parallel to the plane of vibration of the polarizer or analyzer. Symmetrical extinction occurs when the two vibration directions of the
crystal are oriented in a symmetrical manner within the crystal. Oblique
extinction is shown when the long direction of the crystal forms an
oblique angle with the plane of vibration of the polarizer or analyzer.
The extinction angle is the angle formed between the long axis of the
crystal and the plane of vibration of the nearer of the components.

Other optical properties or manifestations of optical properties that
can be determined with a polarizing microscope and can be used to distinguish among crystals of different materials or even identify them are
interference figures, the amount of retardation, the sign of elongation,
and the refractive indices. However, the determination of these properties for each of a large number of particles, to distinguish among them
with regard to composition, is usually much too time consuming to be
practical.

1.7.   Dispersion Staining

Dispersion staining is an optical method for distinguishing among
different types of particles and derives its name by analogy with the
dye staining of biological preparations.[18] It is a phenomenon that
occurs when a transparent, colorless specimen is mounted in a liquid
on a microscope slide and the wavelength versus refractive index (dispersion) curves cross, that is, when the refractive indices of the specimen and mounting liquid are the same for some wavelengths and different for others. The edge of the specimen then appears colored when
viewed through a microscope. By the use of axial illumination and of
stops at the back focal plane of the objective, different colors can be
produced for the same combination of particle and mounting liquid. An
annular stop produces a color made up of wavelengths close to the
wavelength for which the dispersion curves cross, whereas a central stop
produces colors complementary to those shown by the annular stop.
Crossmon[19] has used the method to distinguish corundum particles when
mixed with small amounts of quartz, mica, magnetite, and so on; he
emphasizes that the method can be used for the easy identification of any
one transparent constituent of a mixture of particles that differ substan-

tially, especially in refractive index. He recommended the use of an Abbe condenser of 1.25 N.A. and an achromatic 10× objective of 0.25 N.A. A very bright microscope lamp must be used and should be equipped with a filter to produce nearly white light. Instead of using a stop back of the objective, Crossmon recommended removing the upper element of the Abbe condenser and placing a dark-field stop 16 or 17 mm in diameter in the carrier below the condenser. For corundum, Crossmon used methylene iodide as the dispersing liquid. After focusing the microscope on the preparation, the condenser is moved up or down until dark-field illumination is achieved. Under these conditions the corundum particles are yellow and have purple or blue borders.

For other types of particles Crossmon recommended the selection of an immersion liquid having a refractive index near that of the particle and a much higher dispersion. For an index range of 1.440–1.628 he suggested mixtures of diethylene glycolmonobutyl ether and α-chloronaphthalene. Mixtures of the latter with α-bromonaphthalene were suggested for the range 1.632–1.656, and of α-bromonaphthalene and methylene iodide from 1.660 to 1.70. He also stated that mixtures of diethylene glycolmonobutyl ether with cinnamaldehyde (1.440–1.62) produce stronger colors than the mixtures described above.

Consider a situation in which the refractive index of the liquid is greater than that of the solid for blue light, equal for yellow light (λ_0), and less than that of the solid for red light. Furthermore, the solid object is illuminated with white axial light. Short wavelengths will be refracted toward the liquid, long ones toward the solid, and those near λ_0 little or not at all. This variation in refraction produces the color spectrum at the edge of the particle. Now if a central stop is placed behind the objective lens, the axial beam is stopped, allowing white light minus the wavelengths close to and including λ_0 to pass. If the stop is annular (wide, with a hole in the middle), it allows only light close to and including λ_0 to be transmitted. For the situation just described, the annular stop produces images with yellow edges on a bright field, whereas the central stop produces blue images on a dark field. The stop should be of such size and position as to stop or transmit only the axial beam. McCrone and Delly[18] state that diameters of 3 mm for the annular stop and 4 mm for the central stop are satisfactory with 10× objectives.

Commercially available liquids can be employed as mounting media. The use of Cargille refractive index media* has been described by Grabar.[20] McCrone and Delly provide a table of colors observed for

* R. P. Cargille Laboratories, Inc., Cedar Grove, N.J.

different values of λ_0. From 4300 to 6600 Å the colors vary from blue-violet to brown-red for annular stops and from yellow to pale blue green for central stops.

Cherkasov[21] used somewhat different techniques to produce the dispersion colors. Instead of the annular stop, an iris diaphragm was placed at the back focal plane of the objective and was closed to the appropriate extent. He also used a central stop in place of the iris diaphragm, but this could be used off-center to produce different colors on opposite sides of the image.

If the particle is anisotropic, the dispersion colors will vary with the particle orientation because of the differing refractive index when the illumination is with polarized light. This effect can be observed using one polarizer of a polarizing microscope. With unpolarized light an anisotropic crystal simultaneously displays all the colors it would show with a single polarizer during rotation of the stage.

1.8. Photomicrography

Photomicrographs of particulate material can have several fairly obvious uses related to the measurement of airborne particles. Some methods of particle measurement involve a preliminary preparation of photomicrographs of the collected particles followed by measurements of the images of the particles. The measurements may be manual, semiautomatic, or automatic. Photomicrographs provide an excellent, though qualitative, indication of particle shape and general appearance, and for this reason are often included in the published literature dealing with aerosols and other aspects of fine particle technology. And of course, the photomicrographs provide a permanent record that can be referred to whenever desired.

Any microscope and camera can be used for taking photomicrographs. Even a simple box camera can be used. The only additional equipment that is essential is some means for mounting the camera above the microscope eyepiece. A light trap between the camera lens and the eyepiece to screen out extraneous light is useful but not necessary. Most persons using a microscope automatically focus the microscope with their eyes adjusted for observing objects at great distances. Therefore, usually the light rays emerging from the eyepiece of a focused microscope are axial and parallel, and the camera will form an image on the film if focused for infinity. A fixed-focus camera is designed to focus from some relatively close distance to infinity. Chamot and Mason[5] recommend that the camera be focused for about 25 ft, and that this setting may be varied by experience, depending upon whether the observer is near- or far-sighted.

As mentioned earlier, 35mm cameras are popular for photomicrography because of the large number of photomicrographs that can be taken with a single roll of film. However, such cameras have several disadvantages: (a) unless a roll is completely used, film is wasted or the exposed portion must be removed in a darkroom; (b) the images on the film are very small, and usually must either be enlarged (for prints) or projected; (c) as contrasted with the use of Polaroid film, fairly lengthy development is required. There are advantages. The 35mm single lens reflex cameras equipped for focusing on a disk of ground glass in the field of view have the added advantage that they can be directly focused on the microscope field of view. Furthermore, the image is visible until the shutter is tripped. The use of 35mm cameras also greatly decreases the expense of taking color photomicrographs.

Larger-roll-film cameras can of course also be used, and produce larger contact prints. If a darkroom is available, cameras accepting cut film (or plates) can be used for black and white or color photomicrography, their use overcoming many of the disadvantages listed above. A ground glass plate can usually be inserted in place of the film for focusing.

Many modern cameras designed for photomicrography will accept interchangeable backs for cut film, Polaroid film, or ground-glass plates.

A camera lens is necessary only if the camera itself is to be used for focusing, for example by prefocusing it for infinity. Otherwise the camera lens can be omitted and the image is formed on the photographic emulsion with the ocular of the microscope. Almost any ocular can be used, but flat-field oculars designed for this purpose are available. Obviously, when the microscope ocular is used to form the image, the rays emerging from the ocular are not parallel, and simply focusing with the camera removed through the microscope eyepiece is useless.

Although photomicrographs can be taken with any camera, a shutter equipped with a cable release has the advantages of convenience and of avoiding shaking the camera when lengthy exposure times are required. Rather long exposure times are often needed, especially when working at high magnifications. Shutters providing automatic timing for exposures up to 1 sec are convenient and greatly increase the accuracy of exposures of intermediate length. Rather elaborate devices are also commercially available for automatically controlling exposure times (Figure 3-27).

Vibration of the camera-microscope system can result from an unsteady mounting system for the camera, causing blurring of the photo-

Figure 3-27. System for automatic shutter control and polaroid camera for photomicrography. Courtesy of the American Optical Co.

micrographs, and an unsteady table on which the microscope rests can have a similar result. Even machinery, such as a vacuum pump operating in a nearby room, may produce so much vibration that photomicrography is impossible. The stands of many photographic enlargers are quite rigid and can double as stands for photomicrography. In fact, some photographic enlargers have photomicrographic adapters as accessories. Schaeffer[16] has described the preparation of a small cardboard adapter tube for mounting certain small, light cameras such as many 35mm cameras at the top of a microscope.

Focusing on a ground-glass plate, with the plane of the ground glass surface in that of the photographic emulsion, is the most precise focusing method. Some microscopists find that even sharper focusing can be achieved by having a clear glass area within the plane of the ground glass; they obtain the final focusing with the very faint but potentially very sharp image produced on the clear glass. A clear glass area can be produced by cementing a microscope cover slide onto the surface with Canada balsam. By placing a focusing magnifier on the clear glass screen, a sharp, bright image is obtained when a real image is produced in the plane of the clear glass.

Many commercially available photomicrographic systems include a side tube for viewing and focusing with the camera (and film) in place.

The optics are such that when the particles appear in focus to the observer, an image is focused on the film by the turn of a prism or mirror. The accommodation of the eye and the variation among eyes (near-sightedness, etc.) sometimes lead to difficulties when using this method.

Exposure meters for photomicrography are commercially available and are helpful, but they can give misleading results when a large amount of contrast occurs in the field, as is so often the case during fine particle work. The results can be improved by exposing three frames using the exposure indicated, one shorter, and one longer.

An excellent method for determining the correct exposure when using cut film in film holders or Polaroid film makes use of the slide that protects the film from exposure during handling. Such a slide is part of the film holder for cut film and part of each Polaroid composite. With the film in place in the camera and the shutter closed, the slide is pulled out far enough to uncover a narrow strip of film. The film is then exposed for the longest time that might be appropriate, the slide pulled out again and a shorter exposure time used. This process is repeated until many strips of film have been exposed for decreasing times. Examination of the developed film will usually reveal an appropriate exposure time.

Color photomicrography is useful for producing records and demonstrating certain features. Success depends on more careful control of exposure than when black and white film is used, and the matching of the "color temperature" of the light source with that of the film. Color temperature can be defined as the temperature to which a blackbody must be raised to produce a given color, from deep red to a blue-white. Of course the blackbody emits a wide range of wavelengths for each temperature, and a precise match of blackbody radiation with the color of light used for illumination is seldom possible. Nonetheless, the match can almost always be sufficiently close for photography. Accordingly, color films are generally prepared ("balanced") for use with light of a specified temperature. If a film is balanced for daylight, it can often be used with an incandescent lamp, such as those designed for indoor photography, and an appropriate blue filter. Other films are balanced to be used with such lamps without filters, but the lamp must have the proper color temperature. Information concerning combinations that will give the proper color temperature is often included with the film by the manufacturer. Color temperature meters are commercially available, and the color temperature of incandescent lamps can be progressively changed with a rheostat or a variable transformer.

1.9. Measuring the Particles

The dimensions of the particles are usually measured directly, using ocular micrometers. These are scales that can be placed in the focal plane of the ocular, and many oculars are so designed that the upper lens can be adjusted, if necessary, to bring the scale into sharp focus. The type of micrometer used depends somewhat on the definition of diameter employed. If the diameter is defined in terms of the actual dimensions of the particle, a micrometer that has a linear scale is convenient, but other scales may be preferable if the diameter is defined in other ways, for example, as the diameter of a circle whose area is estimated to be equal to the projected area of the particle. An ocular micrometer (graticule) for making this estimate was designed by Patterson and Cawood[22] for rapidly estimating the size distribution of smoke particles in the range 0.2–5.0 μm diameter. The micrometer consists of a rectangle divided into smaller rectangles and two sets of numbered circles (Figure 3-28). The areas within one series of circles is blackened, giving rise to the name globe and circle for this type of graticule. The microscopist has the choice of using either the globes or the circles for making his estimates. The rectangle is intended to define a precise area within the field of view such that all particles within that area are counted and measured. Fairs[23] designed a series of three graticules de-

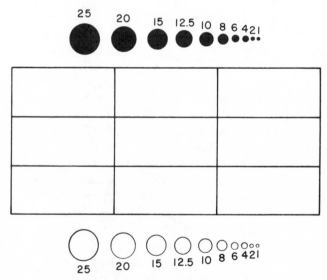

Figure 3-28.   Globe and Circle graticule.[22] Courtesy of the Chemical Society.

signed to cover a greater range of sizes than the Patterson–Cawood graticule. The size intervals except for the two lowest are arranged in a $\sqrt{2}$ progression.

Fairs[23] method of particle sizing required two observers, one making the measurements and the other recording them. He found that two or three magnifications were usually required to include an entire size distribution.

A modification of the globe and circle graticule was developed by May.[24] The circles have been numbered according to their powers of $\sqrt{2}$. May suggested measuring the particles using a mechanical stage, sizing with the globes or circles but making frequent checks with the distance between vertical lines.

Other modifications of the globe and circle graticules have been made by Watson,[25] and by Hamilton et al.[26] These were considerably simplified over previous designs, because experience with the results of using the more detailed graticules showed that much of the information was not used. Hamilton's design was for use by the British National Coal Board. The circles corresponded to 0.5, 1, 2.5, and 5 μm when used with a 2-mm objective and the appropriate tube length and eyepiece. The intermediate value of 2.5 μm was chosen so that the sizes 1 and 5 μm would still be indicated when the magnification was halved when changing from a 2-mm oil-immersion objective to a 4-mm dry objective. It has been adapted for routine dust counting by the National Coal Board and is produced in two sizes with net lengths of 4.0 mm and 3.5 mm to be used with positive and negative types of eyepieces, respectively.

Hand-held test cards have been designed for training personnel in the use of globe and circle techniques.[27] The trainee compares profiles representing particles with the globes or circles drawn on the cards. Results obtained with such cards have indicated a tendency to overestimate the areas within the profiles, and the training helps overcome this bias.

The processes used in the manufacture of various types of micrometers have been described in detail by Bovey.[28]

Eyepiece micrometers consisting of parallel rulings are used for determining Martin's diameter or Feret's diameter. Errors result from the use of such micrometers having fixed scales, because the measurement usually requires an estimation of a fraction of a division. This source of error can be overcome using filar micrometers, although their use is rather cumbersome. Filar micrometer eyepieces are designed for moving a scale or cross hair across the field by means of a screw attached to a graduated drum. The divisions on the drum are calibrated in terms of those on the ruled eyepiece micrometer. Some filar micrometers are designed so that one complete rotation of the drum moves the cross hair one division of the eyepiece micrometer. Other micrometers have a fixed cross hair and moving scale. To measure particles with a moving scale type of micrometer, the drum is set at zero and the slide bearing the particles is moved with the mechanical stage until one edge of a particle just touches zero on the scale. The number of whole divisions of the scale covered by the particle diameter is then counted and the scale moved by means of the graduated drum to bring the last division mark covered by the particle to the end of the diameter remote from the zero line. The reading on the drum is the fraction of a scale division to be added to the whole number of divisions covered by the diameter. The filar micrometer equipped with a movable cross hair is used in an analogous manner.

Another type of eyepiece for measuring particle sizes is based on the principle of image splitting or shearing, which has been used for many years for making astronomical measurements. Two images of each object are formed in the field of view and a means is provided for moving them relative to each other. To make a measurement the images are first made to coincide and then are separated until they just touch. The amount of shearing is then proportional to the diameter of the particle, which can be determined from a calibration of the device. Several such instruments are commercially available, and in one the images are of different colors, red and green.

All types of eyepiece micrometers or graticules should be calibrated for every combination of objective, eyepiece, and tube length employed, using a stage micrometer with linear rulings. The closest spacing of the rulings is usually about 10 μm. With the microscope focused on the rulings, the stage micrometer is moved with the mechanical stage until one of the lines of the stage micrometer precisely coincides with the zero line of the ocular micrometer. The tube length is then adjusted until both the extreme lines of the ocular micrometer coincide with lines on the stage micrometer. The number of stage scale divisions covered by the ocular scale, multiplied by the actual distance between each stage scale division and divided by the number of ocular scale divisions, is the length represented by each ocular scale division.

As demonstrated by Figure 3-3, the measured particle diameter is larger than the actual diameter by an amount approximately equal to the limit of resolution of the objective, although the enlargement varies somewhat with the nature of the illumination and the ability of the microscopist to detect small differences in the intensity of light. Some microscopists subtract the theoretical limit of resolution of the objective from the measured diameters.

1.10. Slide Preparation

Many methods of aerosol particle collection produce deposits of particles on microscope slides that require only the addition of a mounting oil and cover glass to complete the preparation. An example is collection by sedimentation. However, other methods produce samples in a form such that transfer to a microscope slide is required, an example being the collection of fairly large amounts of particulate material by the filtration of stack gases. Thus a brief discussion of slide preparation is warranted here.

If the sample is small, it may be entirely suspended in some dispersing liquid. A drop of the resulting slurry is then transferred to the microscope slide with a small spatula or glass rod. If the particles are large, dense, or both, the dispersing liquid should be sufficiently viscous to prevent appreciable sedimentation. Another procedure is to transfer a very small sample of the dry powder to the slide and disperse it in a drop of liquid placed on top of the sample.

A nearly optically homogeneous path (except for the particles) between the back lens of the condenser and the front lens of the objective is required to achieve maximum resolving power. This homogeneity is in part achieved by using a mounting fluid between the microscope slide and the cover glass. However, if the sample is a powder the mounting fluid can play the additional role of helping disperse the powder evenly on the microscope slide. Usually aggregates should be broken up, but care must be taken not to crush soft or brittle particles. Other requirements are that the particles must not be classified according to size by the dispersing method, that the dispersing liquid does not react physically or chemically with the particles, and that it has a refractive index distinctly different from that of the particles.

Numerous methods have been proposed for dispersing the powder and mounting it beneath the cover glass. Green[29] has suggested dispersing a pigment on the slide in redistilled turpentine, evaporating the turpentine, melting a small piece of dammar placed on the particles, and pressing a cover slide onto the dammar and particles. First[30] recommended using a toothpick to disperse the particles, because the wood is soft and flexible, minimizing shattering. A drop of mounting medium is placed on a clean glass slide and a small quantity of the powder is added from the end of the toothpick. The dust is rubbed into the drop and the suspension is spread over the glass as a thin film. Placing a clean cover glass over the film completes the preparation. Very fragile particles can be dispersed in a few milliliters of a dilute solution of some resin, such as dammar in turpentine, using hand or machine shaking. A drop of the resulting suspension is placed on a microscope slide, covered with a cover glass, and allowed to harden.

Many liquids have been used for dispersing powders, the selection often being made by trial and error. Wetting agents are often dissolved in the liquids to aid the disperson. Canada balsam, which is often used for various purposes in microscope work, is not suitable since its refractive index, 1.53, is close to that of most particles. The Aroclors are especially suitable since their refractive indices are in the range 1.64–1.67.[18] Various household cements can be used for permanent mounting or can be used to seal the edges of the cover glass to the microscope slide. If the particles are to be recovered, a volatile dispersing agent such as 1,4-dioxane, toluene, or methylene iodide is used.

Particles collected on membrane filters can be mounted by placing a portion of the filter on a microscope slide and adding a liquid having a refractive index close to that of the filter. Appropriate liquids are recommended by the manufacturers. The filter is rendered transparent, and the preparation can be completed by adding a cover glass. Particles that are larger than the pores of the filter will have been collected on the upper surface of the filter and thus will be in a single plane.

Another method of mounting involves dissolving the filter in a solvent for the filter material and mounting a portion of the resulting suspension of particles, or separating the particles from the suspension by centrifuging. The separated particles can be redispersed in a mounting medium as described above. This approach can, of course, be applied to soluble fiber filters, such as polystyrene fiber filters, as well as to membrane filters.

1.11.   Selection of Fields

In order to obtain statistically significant results, usually at least 200 particles are measured. Airborne particles usually have a very wide size distribution, with large numbers of the smaller particles for each of the larger ones. To obtain statistically significant results for the large particles, an exorbitant number of small ones must be measured unless the sample is broken down into various size classes which are studied separately. The multistage impactor, which physically classifies the particles, was developed to satisfy this requirement. However, physical separation is not necessary, and the following method or some modification of it can be used.

A convenient number of class intervals and the class limits are chosen, based on a preliminary examination of the sample to form an estimate of the width of the size distribution, or based on previous experience with that type of sample. The microscope is focused with the highest powered objective to be used, and all the particles in the field are counted in each of the three or four class intervals containing the smallest diameters. If an ocular micrometer such as that shown in Figure 3-28 is used, the class limits can be set by the size corresponding to the globes or circles, and the particles within one or more of the rectangles rather than within the entire field of view can be measured. This procedure is repeated for several fields until about 100 particles have been counted. The count is repeated with a lower-powered objective in three or four class intervals containing larger diameters. The ocular micrometer must, of course, be recalibrated for each magnification, and the magnification adjusted, if globe and circle micrometers are used, so that the class intervals are precisely adjacent to those at the higher magnification. About 50 particles are measured and counted, and the procedure is repeated with a still-lower-powered objective, counting about 50 particles in the remaining class intervals. The area in the field or rectangle for each objective is measured, and the results are combined by recalculating all the data in terms of a single area.

The selection of area (fields of view) where the particles are to be measured and counted can be random or according to some predetermined pattern, but whichever method is used, the areas should be selected without observing them through the microscope. Otherwise, unconscious preferential selection of fields may occur. Fields chosen according to a pattern can be laid out in terms of the rulings on the mechanical stage.

1.12.   Measurements on Photomicrographs and on Projected Images

Many microscopists prefer to measure photographic or projected images of particles. The images on negatives or prints can be measured with any convenient rule. Transparent rules are especially useful, and they can be in the form of the globe and circle graticules rather than linear rulings. If the photomicrographs are on 35mm film, enlarged prints will generally be required. The images can be checked off as they are measured to ensure that all the images are measured and that no duplication occurs. Similarly, the images can be projected onto a screen, either directly, using a projection eyepiece and a very intense light source, or indirectly, first preparing photomicrographs which are then projected. The screen can be a sheet of paper, the images checked off as they are measured. Most observers find that such methods are less tiring than making direct measurements through the microscope, which may require hours of measurement and observation.

1.13.   Particle Thickness

The thickness of particles that are nearly equidimensional or are merely elongated can be inferred from measurements made as described above. However, the thickness of many particles, such as plates, can be obtained only by direct measurement. When the thickness of a particle is much greater than the depth of field of the objective, the thickness can be determined by focusing on the bottom of the particle, noting the reading on the graduated fine focusing adjustment, and then focusing on the top of the particle, again noting the reading. The graduations on the fine adjustment must be calibrated in terms of distance of movement of the objective per given number of graduations. This can be accomplished with a thin glass rod or uniform fiber placed on a microscope slide. The width of the rod is measured using an ocular micrometer, and the thickness is then measured in terms of the fine graduations, focusing on the bottom of the rod and then on the top. The focusing movement should always be in the same direction; otherwise backlash in the gears may introduce errors. These measurements should be repeated several times. The measured rod or fiber width, which is also its thickness, divided by the rod thickness in terms of graduations provides the desired calibration. Because the depth of field of objectives decreases with increasing numerical aperture, objectives having a large numerical aperture should usually be used for such measurements.

An excellent method for measuring particle thickness when a large number of particles must be measured is the shadowing technique used extensively in electron microscopy, described in detail under that heading. Essentially it consists of electrically heating a small amount of a metal, usually in thin sheet or wire form, in an evacuated chamber which also contains the bare particles on a slide. The vapor of the metal approaches the slide obliquely as a nearly parallel molecular beam and

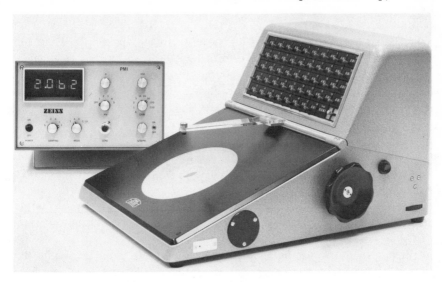

Figure 3-29. Semiautomatic particle sizer with photometer indicator. Courtesy of Karl Zeiss, Inc., New York.

deposits on the slide except where it is shaded by the particles. If the beam forms a 45° angle with the slide, the length of the shadow of a particle equals its thickness. Shadowing units are commercially available, designed primarily for electron microscopy.

1.14. Semi-automatic and Automatic Measuring

Numerous semi-automatic and automatic counting and measuring systems have been developed to remove much of the drudgery of using the methods described above. One of the simplest of these devices is manufactured by the Carl Zeiss Company (Figure 3-29). A photomicrograph or electron micrograph is prepared and is placed on a plastic plate on the instrument. An image of an iris diaphragm is produced in the plane of the plate, appearing as a bright spot that can be superimposed on each image of the photomicrograph in turn. The iris diaphragm is adjusted until, in the judgment of the operator, the area of the bright spot equals the area of the image of the particle. The diameter of the spot is then registered as a single count in an appropriate size channel by depressing a switch. Because there are 48 size channels, a detailed size distribution is obtained.

Another semi-automatic device is the Timbrell Double Image Micrometer, based on the image-shearing (double image) principle. Counting
rates as high as 600 particles in 10 min have been claimed. The shearing
is provided by a chopped light source and vibrating mirror, and the
energy supplied to the vibrator is calibrated in terms of the shearing distance. The separation of the two images is adjusted to a fixed value, and
a foot pedal is pressed to record each particle in that size group on a
counter.

Much more completely automatic are the numerous flying-spot counters and particle-measuring instruments which have been designed in a
wide range of sophistication. The flying-spot counters start with one of
the systems described earlier for producing visible images by ultraviolet
microscopy. The light source can be a flying-spot scanner tube on which
a rastor is traced by the spot. The microscope is then used in reverse,
an image of the rastor being superimposed on the particles. Light that
is not absorbed by the particles strikes the active surface of a photomultiplier tube beneath the particles. The resulting current is amplified
and modulates a monitor tube synchronous with the flying-spot scanner
tube, producing a greatly magnified image of the field of view. Because
an image is produced, obviously all the information is carried by the
current from the photomultiplier which is needed to yield a particle
size distribution, and various methods have been used, or proposed, for
deriving the distribution. Numerous techniques for accomplishing this
derivation were discussed at a session of the Conference on the Physics
of Particle Size Analysis held at the University of Nottingham in 1954.[31]
One difficulty is correction for particles overlapping the field of view.
Another difficulty arises from the fact that the spot of light will traverse
a large particle many times, and the computer analyzer of the signal
must allow for this. An example of such a system is that described by
Roberts and Young,[32] which has been commercially available from Cin-
ema Television, Ltd., of London. It has been criticized on the grounds
that the light transmitted may show some variation when the beam is
in an oblique position and that there are imperfections in the optical
system. However, this will not interfere where only on/off signals are
required, as in the commercial instrument. The sizing is accomplished
by pulse-width selection, and all pulses above a predetermined size are
recorded. Ames et al.[33] compared the size distributions obtained with
this instrument with those obtained by sedimentation for a number of
different powders. Quite good agreement was obtained.

As in ultraviolet microscopy, the microscope can be used in the normal
manner, the image being produced in a television camera placed at the

eyepoint of the microscope. The signal from the camera can be analyzed in many ways, including producing an image. A small computer used at the National Center for Atmospheric Research partially analyzes such signals coming from a television camera over a microscope, and puts the output on magnetic tape. This tape is then fed into a large computer, such as the Control Data 7600, and the data are analyzed to produce particle size distributions in terms of various diameter definitions, such as the diameter of the circle of equal area.

A much more flexible approach is the Bausch and Lomb Quantitative Metallurgical System (QMS). It is an electro-optical system for the automatic measurement, classification, and evaluation of selected features of an image. It was mainly designed for the metals industry for determining properties of metals such as grain size, inclusions, size distributions, percentage area, average length, and Ferets diameter, but it can also be used for automatically determining the size distributions of particles collected from the atmosphere. The heart of the system is a small computer; to change measurement the program is changed. The system consists of a teletype, a system control, the QMS proper, and the teletype interface. The system control receives commands from the teletype and controls the sequence functions of the QMS. Data with the required statistical evaluation are printed out on the teletype in the form of histograms, size distributions, and so forth. The QMS provides automatic microscope stage movement and control, a very useful feature for particle measurements.

The system accepts images from optical microscopes, transmission or scanning electron microscopes, photomicrographs and electron micrographs whether transparent or opaque, and even directly from large specimens. The image is scanned with a highly stabilized Vidicon or Plumbicon scanner. The QMS accepts any standard video output without interfacing. The area measured, which for microscopy is the field of view, and the results of all measurements are displayed on the cathode ray tube of the display unit. The measurement and processing is controlled by various modules. For instance, threshold circuitry distinguishes between the feature of interest and its background. There are four modes of threshold setting: automatic, semi-automatic, manual, and dual-threshold control. In the last mode the system can simultaneously select and measure three gray phases. A wide range of modules is available to perform various measuring functions.

The small computer (system processor) is a high-speed, 16-bit data processor with 8192 words of memory. Sufficient space is provided to contain the programs that control the automation systems, the BASIC interoperative compiler, and up to 300 lines of the computer language, BASIC. The control computation memory is expandable in 4096-word steps to a maximum of 32,768 words. Numerous programs are available from the manufacturer, and the operator can prepare his own.

Measurement frame guardlines are used in conjunction with a "feature specific point" in a way that eliminates any count errors due to double counting. This feature is especially important when counting particles in immediately adjacent fields.

An unusual feature of the system is the light pen, which permits the operator to measure specific features using the selective particle measurement module. A crescent-shaped pattern follows the pen as it is moved across the display. When the pattern is brought to rest on the desired feature, the outline of that feature is formed. Pressing the pen records the result, and measure figures will appear in the usual measurement register of the display. Measurements can be made at the rate of one per second of Feret's diameter, projected length, area including holes, area excluding holes, and longest dimension.

Another interesting feature is two joystick controls that permit the operator to isolate discrete features on the display for analysis.

In addition to obtaining size distributions, the system can perform a statistical analysis of the data, yielding values for statistics such as variance, kurtosis, and skewness.

The measurement by the QMS of images generated by a scanning electron microscope involves some unusual problems. This is because the measurement and analysis system must operate faster than the rate of image generation by the electron beam instrument. This problem is solved in the QMS by interfacing it with a "direct image storage module." The entire image produced by one complete scan of the microscope is placed in the "read" mode of the module and is presented at a high scan rate for measurement and analysis. The transmission electron microscope is interfaced with the QMS by placing a high-aperture lens in the photographic part of the microscope. The lens produces an image on a high-sensitivity phosphor.

Another very elaborate system is the Quantimet, manufactured by Image Analyzing Computers, Ltd.[51,52] The latest version is the Quantimet 720. Like the QMS, it consists of a system of mutually compatible modules which perform various image analysis functions such that various combinations of modules can be combined to perform various analyses. The basic image analyses include area percentages, intersect counts, feature (particle) counts, size distributions, shape factors, and grey values. The system analyzes the images in about 1 sec.

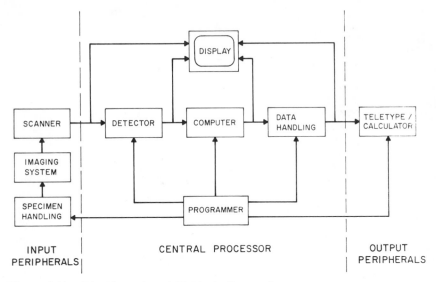

Figure 3-30.   The Quantimet 720 block diagram.[51]

Figure 3-30 is a block diagram of the system. According to Fisher[51] it will usually have an imaging input peripheral, a control processor, and perhaps one or more standard computer output peripherals. The central processor operations can be monitored on the central process display.

Two microscope systems are available and like the QMS include automatic stage control, but other microscopes can be used. Images of electron microscopes can be analyzed by coupling a scanner to the microscope.

The scanner used on the older Quantimet B was a conventional television camera, but such conventional equipment was found to have a number of weaknesses, and a special scanner was developed for the 720.

Edge errors are eliminated in the 720 using two concentric image-limiting frames. The measurements are made within the inner frame, and when the region between the two frames is larger than the largest feature to be measured, no feature that intercepts the outer frame is measured. However, if a feature is within the inner frame, all of it must be inside the outer one and the feature is included in the measurements.

In addition to being presented on the display screen, the results can be fed to a tape punch, teletypes, graph plotters, and the like.

Jesse[52] has recently reviewed quantitative image analysis in microscopy. In addition to the QMS and the Quantimet, he mentions the

Microscan or πMC (Millipore), the Classimat (Leitz), the Microvideomat (Zeiss), and the Telecounter (Schaefer). A very large bibliography is included.

Much less elaborate systems for automatic particle counting and sizing have been developed that use mechanical instead of electronic scanning. The scanning can be accomplished with a fixed slit or spot of light and a moving stage carrying the sample, or the slit or spot can be moved mechanically. Photomicrographs and electron micrographs can be analyzed in a similar manner. When a slit is used rather than a slot, the slit scan sweeps out a path of defined area. Some particles will fall entirely within the scan, others will partially overlap the edge, and at times two or more particles will be intercepted simultaneously. With neither the moving spot nor the moving slit technique does the number of signals give a direct measure of the number of particles. Neither does the length of the intercepts nor the amplitude of the signal give a particle size distribution directly. In using the scanning slit, partial interception of particles at the ends of a slit poses a problem. Statistical and other methods for overcoming these difficulties are discussed in the various papers of the symposium mentioned above.[31]

Casella Electronic, Ltd., has manufactured an instrument for counting and measuring particles using a stationary slit and a moving stage. The transmitted light intensity is measured with a photomultiplier tube, and the length of slit intercepted by the particles is used to obtain particle size distributions. Corrections for the partial interception of particles at the ends of the scanning slit are made by scanning with two slits of different lengths.

1.15.   Particulate Tracers

So long as air pollution cannot be completely eliminated, questions will arise concerning its transport. For example, many companies have been sued for supposedly damaging crops with effluents from their stacks. Such a company may want to counter by determining the extent, if any, to which damaging concentrations of stack gases reach the crops. Another example is research on the dispersion of stack gases under different meteorological conditions. Although many of the questions can be answered by meteorological studies alone, the determination of details of the transport often requires that some readily identified tracer be introduced into the pollutant or the polluted air so that its motions can be followed and the concentrations at various distances downwind determined. Sometimes one of the pollutants can be used as the tracer, but this is not practical if there is more than one source, either natural or anthropogenic.

Until recently the tracers have almost always been particles. Particulate matter is still used, but a gaseous tracer, sulfur hexafluoride, is also becoming popular because it is chemically inert and can be detected at very low concentrations by the electron capture detectors developed for gas chromatography. The chief advantage to the use of particles is that a single unique particle suspended in a very large volume of gas can often be detected, whereas most gaseous tracers may become so diluted that the determination of their concentrations is impossible with reasonable techniques. Of course, the particles must be sufficiently small that sedimentation is negligible, but large enough so that they can easily be detected with a microscope. The diameter range 0.5–5 μm is often employed.

The most widely used particulate matter has consisted of inorganic fluorescent substances such as zinc sulfide, zinc silicate, and zinc cadmium sulfide, especially the latter. Water-soluble fluorescent dyes such as uranine have also been used. Fluorescent particles have been employed so extensively as tracers because they are easily distinguished from naturally occurring particles by color and brightness. Fluorescent particle concentrations as low as one in 10 ft³ of air can be measured with a sample size of about 500 ft³.

Several methods have been used for dispersing the particles in the air stream. For example, Braham et al.[34] dispersed a synthetic crystalline zinc sulfide, manufactured by the New Jersey Zinc Co., which fluoresces green. The fluorescence is most intense with illumination at 3650 Å. Two dispersal techniques were used. Usually the pigment was mixed with an oil and fed into a military smoke generator. The other technique utilized a disperser consisting of an 8-in. rotor spinning at 10,000 rpm with its flat top separated from a stationary plate by 0.030 in. Air and pigment were admitted to the space between the plates near the center of rotation and discharged near the edge. Another method involved eroding compressed pellets of the tracer material with a jet of compressed air.

Braham et al. sampled from an airplane, collecting the particles by impaction on a glycerol-coated strip of transparent plastic. The strip, in cylindrical form, was held on a disk rotated by clockwork so that the strip passed in front of the impacting jet at a speed of 1/8 in. per minute. More commonly, the particles are collected on membrane filters having pores sufficiently small that essentially all the particles are collected on the filter surface, in a single plane.

The numbers of fluorescent particles in the volumes of air sampled are determined by counting them with a microscope while illuminating them with ultraviolet light. A low-power, high-dry objective can be used and the ultraviolet light shined directly on the field by the ultraviolet lamp. Shielding must be provided so that the eyes of the person counting are not exposed to either direct or scattered ultraviolet radiation. Statistical methods for counting particles have been described that permit accurate control of the probable error.[35] A rapid preliminary count is made of the particle density on each filter to provide data for estimating the number of fields to be counted to provide the desired accuracy.

The use of water-soluble fluorescent dyes avoids the tedious and expensive particle counting.[36] Aqueous solutions of the dye are dispersed, for example by aspirating-type aerosol generators, and the droplets evaporate to yield the desired particles. Fiber filters can be used for the sampling, and at much greater velocities than is usually reasonable with the membrane filters, which have a high flow resistance. The filters are extracted with water and the dye content determined with a fluorescence photometer. Such dyes cannot be used in hot stack gases where they would be destroyed.

2. TRANSMISSION ELECTRON MICROSCOPY

2.1. Electron Optics

Electron microscopes, such as the Philips shown in Figure 3-31, have many similarities to optical microscopes, with electrons replacing photons. Any moving body behaves as though it has a wave associated with it, and the wavelengths associated with electrons are extremely short. Thus extremely high resolving power is possible for electron microscopy as compared with optical microscopy. The wavelength associated with any moving body was shown by deBroglie to be given by the equation

$$\lambda = \frac{h}{mv} \tag{3-4}$$

where h is Planck's constant, m is the mass of the body, and v is its velocity. The wavelength associated with electrons which have been accelerated through a potential difference V is given by the equation:

$$\lambda = \frac{12.3}{\sqrt{V}} \tag{3-5}$$

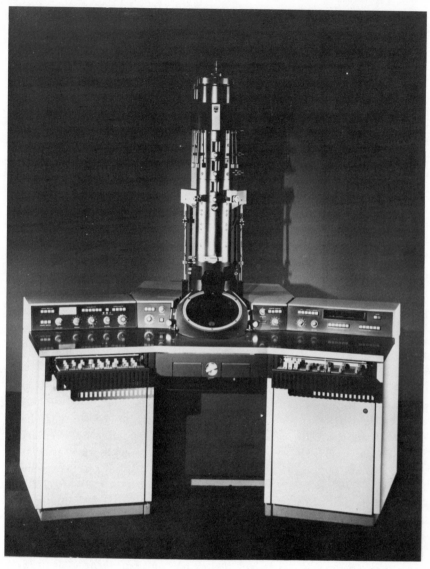

Figure 3-31. Philips EM 301 transmission electron microscope. Courtesy of Philips Electronic Instruments, Inc.

where V is in volts and λ is in angstroms. The wavelength of an electron beam may be a small fraction of an angstrom, whereas the wavelengths for visible light are in the range of thousands of angstroms.

An electron beam is completely stopped by rather thin layers of solids and thus cannot be focused by refraction on passing through a solid as can a light beam, and the path of the electron beam in an electron microscope must be evacuated. Therefore the beam is focused by magnetic or electrostatic fields produced by what are in effect lenses for the beam. The major lenses of an optical microscope have their counterparts in electron microscopes. Thus there is a condenser lens that focuses the electron beam onto the object, an objective lens, and a projector lens which is the counterpart of the ocular.

The lenses of electron microscopes suffer from both spherical and chromatic aberration. The former, previously defined as the failure of a lens to bring rays of a single wavelength to a single focal point (Figure 3-6) is partially avoided by making the aperture angle of the objective small. However, this decreases the numerical aperture and thus the resolving power of the microscope. Therefore the design of an electron microscope includes a compromise between resolving power and spherical aberration. There is one compensation, however, for this loss of resolving power. The depth of field of an objective can be calculated from the expression $c/\tan$ (A.A./2) where A.A. is the angular aperture and c is the diameter of the circle of confusion one is willing to accept. This diameter is often chosen to equal the limit of resolution and sometimes twice the limit of resolution. Because of the small angular aperture, the depth of field is usually two or more orders of magnitude greater than the limit of resolution of the electron microscope. The large depth of field is usually an advantage, although it does prevent optical sectioning of a particle whereby structure at several depths through a specimen can be observed with an optical microscope by changing the level in focus. However, having the entire particle resolved is a great advantage when measuring large numbers of particles.

The lenses of electron microscopes also produce various distortions. Magnetic lenses may produce isotropic distortion, which is observed as pincushion distortion or barrel distortion, and anisotropic distortion, which produces a pocket handkerchief shape. These distortions can be especially severe for intermediate and projector lenses, described later. Distortion coefficients are usually furnished by the manufacturers.

The electrons of the electron beam are produced by an electron gun, usually consisting of a tungsten wire cathode, an aperture-grid (Wehnelt cylinder), and an anode. The cathode is V shaped, so that when an electric current is passed through the wire, the apex of the V is the hottest part and thus is nearly a point source for the thermal emission of electrons. The anode is maintained at ground potential and the cathode at the accelerating potential, usually 50–100 kv. The grid (guard cylinder, or shield) can be unbiased with respect to the cathode or operated with a negative bias of a few hundred volts. The biased gun gives a more intense beam, whereas the unbiased gun is used when it is desired to make the specimen a virtual source. The bias can be achieved by connecting the cathode to the negative terminal of a high-voltage power supply through a resistor and connecting the grid directly to the negative terminal. This method of producing bias is self-stabilizing. When the cathode is at room temperature, no electrons are emitted, and since no current flows through the resistor there is no bias. As the cathode temperature is increased, the bias voltage increases until it is sufficiently large that some emitted electrons are expelled back toward the cathode. A further temperature increase produces little additional beam intensity. In order to obtain a reasonably intense image, the cathode-source must be operated with such a high current that rather rapid evaporation of the tungsten occurs and the filament only survives for 10 or a few 10s of hours. Of course, an electron microscope is often not operated at the maximum beam intensity. Many types of particles, such as those consisting of ammonium salts, decompose or evaporate in the electron beam, and a very dim image must be tolerated to prevent the rapid disappearance of the particles.

The electron lenses produce nonuniform magnetic or electrostatic fields that have been described as being bell shaped along the axis of the microscope. There is a close similarity between the behavior of a photon passing through a glass lens and of an electron passing through the field of an electrostatic lens. In each case both direction and speed are altered within the lens, and the paths through the lenses are similar. However, the path of an electron through a magnetic lens is much more complicated. Direction of the electron but not its speed is changed as it travels through the magnetic field and the paths are helical.

The electrostatic lens consists of three parallel plates having axial holes. These plates are the electrodes, the outer ones being at ground potential and the inner one at the potential of the filament in the electron gun. The resulting field is essentially lens shaped, extending outward from the axial holes. Electrostatic lenses are seldom used today for image formation.

The magnetic lens consists of a solenoidal excitation coil completely enclosed in an iron case, except for an interior gap, and an axial hole to permit the passage of electrons. Pole pieces may be inserted to further concentrate the field.

The condenser lens, as in a light microscope, focuses the rays onto the particles, but in addition is used to regulate the intensity, which can be controlled by the lens in two ways. One is by controlling the lens aperture, a larger aperture permitting more electrons to reach the object than a small one and thus producing a greater image intensity. The other method is to change the focus of the lens by changing the field strength so that the electron rays are brought together either above or below the particles. Since such defocusing spreads the beam over a larger area, the intensity is reduced. Such methods can, of course, also be used to control the intensity in light microscopy, but are not the most desirable. Fischer[37] suggests that the aperture angle of the condenser should be kept small for maximum depth of field and possibly also resolution. This can be accomplished with a small condenser lens aperture or by adjusting the strength of the condenser lens field to spread out the electron beam at the field. Both these approaches, as just mentioned, decrease the image brightness, so it would seem that a compromise is required. Fischer points out, however, that a relatively bright image is needed only for visual observation, not for photographic reproduction of the image, so an excessive angle of convergence can be used for visual observation and focusing, and the angle reduced for photography.

Modern electron microscopes are equipped with two condenser lenses for greater flexibility. The electrons can be concentrated on a very small area so that the final image is bright even at very high magnifications.

Just as the objective lens of a light microscope is the most critical part of its optical system, so is the objective lens of the electron microscope of controlling importance since it determines the achievable limit of resolution. Electron microscope objectives and projection lenses have high chromatic aberrations in addition to the spherical aberrations already discussed. The lenses cannot be corrected for chromatic aberration which consists of the lens having different focal points for different electron velocities; thus maintaining a constant electron velocity except for changes produced by the particles observed is imperative. The most important causes of electron velocity changes are fluctuations of the accelerating potential and deposits of dirt near the path of the electron beam, especially on the apertures. Therefore, a very constant voltage and periodic cleaning are required. An electron microscope is usually focused by adjusting the field strength of the objective. Thus electron microscopes have coarse and fine objective lens controls.

The objective is often equipped with a separate aperture that corresponds to a diaphragm of a light microscope. The main reason for using an aperture is to improve contrast, since the angular aperture of the objective is controlled by the angle of the cone of illuminating rays. The main disadvantage to the use of such apertures is that they tend to collect dirt, which becomes charged and affects the electron beam.

The objective of many electron microscopes includes a device (a stigmator) to overcome astigmatism otherwise produced by the lens. Astigmatism consists of a point appearing in the image as two perpendicular lines, and results from slight ellipticity in the bore of the lens and from magnetic inhomogeneity in the magnetic material.

The projector lens, like the ocular of the light microscope, produces an image from that formed by the objective. Since the intermediate image is huge compared with the size of the object observed, a large proportion of the rays of electrons entering the projector lens form a relatively large angle with the microscope axis. Spherical aberration results, and magnification increases toward the edges of the field. Such distortions are, of course, especially objectionable when electron microscopy is used for determining particle size distributions. Fortunately, it is most severe when working at low magnifications; at high magnifications only a small portion of the image formed by the projector lens is observed, and particle size measurements are usually made at quite high magnifications. Generally, the magnification by the objective lens is fixed, and the overall magnification is changed by the projector lens. The upper and lower magnification limits that can be attained with a given projector lens can be selected by choosing the appropriate pole piece assemblies. A better approach is to use a third lens between the objective and the projector lens. Such a removable intermediate lens greatly increases the magnification range for a given pole piece. For example, the range of the Zeiss EM10 is $100\times$ to $200,000\times$ with a resolution of 3.5 Å. The use of two lenses can also greatly decrease the distortions resulting from lens aberrations. Kynaston and Mulvey[38] point out that with two or more projector lenses the same magnification can often be obtained by various combinations of the excitations of the projectors. Some of the combinations produce considerable distortion of the final image. They state that the condition for zero distortion does not depend on the excitation of the second projector, and that once the first projector has been correctly set for zero distortion, further changes in the magnification should be made with the second projector. This procedure is especially useful at low magnifications such as $500-3000\times$.

The main power requirements for an electron microscope are (1) the current for the filament in the electron gun, (2) the high voltage for accelerating the electrons, and (3) the power requirements of the lenses. There are, of course, additional requirements for the vacuum pumps and for the pressure (vacuum) meters. Many electron microscopes are now equipped with all-solid-state circuity. The filament is usually heated with a high-frequency alternating current. The accelerating potential is high for three reasons. As demonstrated by equation 3-5, a high voltage is required to produce a small wavelength and thus achieve a high resolving power. A high potential renders inconsequential small variations of the speed of electrons as they leave the electron gun. And the high voltage accelerates the electrons to such an extent that they can penetrate the supporting films for the particles or other specimens and to some extent the specimens themselves. The accelerating potential can be adjusted in some electron microscopes, usually to permit a change in penetrating power and thus a change in contrast. A very large penetrating power can produce an objectionable loss of contrast when observing very thin specimens. The objectives of a few electron microscopes have permanent magnetic lenses, and these must be focused by carefully varying the accelerating voltage. The high voltage is usually produced by rectifying a high-frequency alternating current. The voltage has a ripple, and the variation in the focal length of the objective due to this ripple must be kept less than one quarter of the depth of focus of the instrument.

Although most electron microscopes are designed to operate with accelerating voltages in the range 50 to 100 kV, a few low-voltage microscopes have been built. For example, Wilska[39] has built an instrument with a working range from 1 to 50 kV. It was quite small, the total length of the column being 55 cm and the total weight, including the console, only 162 kg. The voltage regulation occurred through a high-resistance bridge, and the high voltage was changed by means of a number of taps. The resolving power at 25 kV was about 20 to 25 Å and remained much the same down to about 10 kV. Furthermore, electron microscopes with accelerating voltages as high as 3000 kV have been built to examine thick specimens.[40]

Rather conventional regulated power supplies commonly furnish the lens currents. The supply for each lens must, of course, be adjustable.

The optical system of the microscope must be evacuated to a pressure no greater than about 1×10^{-4} torr. Such a vacuum is required to prevent sparking between the anode and cathode of the electron gun and to decrease to an acceptable extent scattering of the electron beam by gas (air) molecules. Also, a low pressure considerably extends the life of the electron gun filament. The pumping system generally consists of a mechanical forepump and a diffusion pump. Some microscopes have been equipped with mercury diffusion pumps which require the use of a liquid nitrogen or dry-ice acetone cold trap. However, the requirement of a cold trap is a nuisance, and oil diffusion pumps that do not require a cold trap are more common. Some microscopes employ three pumps. For example, the Siemens Elmiscope 101 uses an oil diffusion pump, a mercury jet pump, a buffer tank, and a mechanical forepump. The pressures are generally indicated by thermocouple gauges, ion gauges, or both.

The final images are either made immediately visible by forming them with electrons on a fluorescent screen or they are photographed. Both films and plates are commonly used for photography, the latter having the advantage that they remain flat and undergo no distortion during development and handling. Some of the newer microscopes are designed to accept camera systems employing a number of different film sizes, for example, 35mm, 70mm and $3\frac{1}{4} \times 4$ in. The photographic latent images are usually produced by allowing the electrons to fall directly on the photographic emulsion, but in some instruments the fluorescent screen is photographed.

Usually the photographs (electron micrographs) are enlarged, and if the enlargement is excessive, the resulting print will appear grainy. Whether the electron micrograph is enlarged, if the record obtained is to include all of the detail provided by the electron microscope, the magnification to the negative must be sufficiently great that the resolution on the electron micrograph is limited by the resolution of the microscope and not by the grain size of the photographic emulsion.

Exposure times are usually determined by trial and error, as for light microscopy, but exposure meters are sometimes used. Photographic emulsions having wide latitude, that is, having a small sensitivity to exposure time, are desirable for electron micrography because a precise exposure time is not excessively critical. Unfortunately, a high contrast film has a relatively narrow latitude. When an image has little inherent contrast, and high-contrast film is desirable, the exposure time becomes critical.

The specimen, usually mounted on a 3-mm diameter screen, is introduced into the microscope on a narrow stage between the objective and condenser lens, and often within the magnetic field of the objective. The screen fits snugly into a circular compartment so that it is properly aligned. Some stages can be tilted to provide stereo pairs, such as the stage on the Zeiss EM9S-2 microscope. Airlocks are provided to limit admitted air to a small portion of the microscope when a sample is introduced. The specimens can be traversed by means of a servo mechanism similar to that of the mechanical stage of a light microscope. However, the total area of the screen that can be traversed is small and limited to the central portion. Therefore, particulate material should be uniformly spread over the screen or at least be included in the central portion of the screen if particle counts are to be made. This consideration is especially important when airborne particles are collected by slit impactors, and care must be taken that the streak of collected particulate material crosses the center of the screen.

Hot stages have also been produced that can be used to determine the melting points of particles. Two methods for heating are usually used. The simpler is the grid heater in which the specimen support grid is heated by passing a current through it. Furnace-type heaters are also built which provide a more uniform temperature but have a slow response time because of the relatively large heat capacity.

2.2. Particle Mounting

Particles to be examined by electron microscopy are almost always mounted on a thin, fragile, membrane supported by a small circular screen (grid). They are either woven wire screens of about 200 meshes per inch or electro-formed screens having the same mesh density. This mesh has been found to provide adequate support for the mounting films without greatly blocking the field of view. Some early microscopes employed a membrane stretched over a single hole in a metal disk. The grid material is usually either nickel or copper.

The film must be transparent to the electron beam, although it is not invisible, a fact that becomes apparent when the film is broken. Many film materials have been used, but Parlodion and Formvar (polyvinyl formal) are popular because of the strength of the films and the relative ease of preparation. Films of evaporated carbon, various metals such as beryllium or aluminum, and various oxides such as silicon oxide or dioxide are often used.

A number of methods have been used for preparing membranes of Parlodion and other plastics. When Parlodion is used, a common method is to place one or two drops of a 1–2% solution in amyl acetate on a clean water surface. A convenient container for the water is an evaporating dish sufficiently large that the Parlodion solution does not cover the entire surface. The solution quickly dries, leaving the desired thin, tough film. Several clean specimen screens are then placed with forceps, convex surfaces down, on the film. A length of wire is bent into a loop with a handle, the diameter of the loop being greater than the length of a microscope slide. The loop is slipped under the membrane and screens, and lifted from the water. The loop is then inverted and lowered over a clean microscope slide, leaving the membrane-covered screens mounted on the slide. Excess membrane can be folded under the slide or cut away. Each screen and its covering membrane is readily removed from the microscope slide with a pair of forceps, leaving the rest of the film and the other screens attached to the slide. Another method begins as before, by forming a film on water and placing screens on the film. A microscope slide is placed on the film on the water surface so as to cover the screens. The film is cut around the slide and folded back over it. The slide with the attached film and screens is then lifted from the water. A third method involves forming the film on glass instead of on water. The slide is coated with the Parlodion solution either by dipping the slide into the solution or by placing a few drops on the slide surface. The slide is then held in a vertical position to permit it to drain, and is kept in that position until the film is dry. The film is then scored with a needle around the edges and crosswise in order to divide the film into areas of appropriate size. The slide is lowered, in a vertical position, into a tray of water; after about a minute the slide is slowly withdrawn and then slowly resubmerged while the film is floated off onto the surface of the water.[41] If the film does not remain on the surface as it separates from the slide, it can be "teased" to the surface with forceps. The screens are laid on top of the film, and the preparation is completed by one of the previous methods. Fischer[37] states that when Parlodion membranes are prepared in this manner, the solution in amyl acetate should be more dilute, about $\frac{1}{2}\%$ or less, to ensure that the film produced is not too thick. Instead of using a loop to remove the film and grids, Wyckoff[41] suggests placing the grids on pieces of 50-mesh brass screen that are submerged in the water and drawn up under the film.

Formvar films are prepared by coating them, by the method just described, on glass slides, since the solutions of Formvar (in dioxane or in ethylene dichloride) are heavier than water. Concentrations of Formvar between 0.1 and 0.5% are recommended.

If the films cast on glass do not separate readily from the glass when the slides are submerged in water, the separation can often be started with a razor blade, needle, or a bit of Scotch tape.

Films consisting of metals, carbon, and various oxides can be produced by evaporating them in a vacuum and allowing the vapor to condense on a soluble base. The base is then dissolved, leaving the desired film. An advantage to such films is that they are much stronger than those of Formvar or Parlodion.

The most satisfactory method for mounting airborne particles on membrane-covered grids is to collect the particles directly on the grids. When the collections are to be made on microscope slides, the grids and films can be mounted on the slides before sampling, using one of the methods described above. For example, when a slit type of impactor such as the Unico is used to collect particles on a microscope slide, three grids can be mounted across the slide to receive the impacted particles. The impactor surfaces in the immediate vicinity of the jet outlet may have to be cut down to accommodate the grids, and care must be taken to ensure that the modification does not produce leaking. Thermal precipitation and sedimentation are other appropriate methods for collecting aerosol particles directly on membrane-covered electron microscope screens.

Occasionally aerosol particles are collected as dry powder, for example when sampling stack gases with a thimble-type filter. The usual method for mounting such a powder is to prepare a slurry of it in water, place a drop of the slurry on a membrane-covered screen, and allow the water to evaporate. Two major disadvantages to this approach are that the particles tend to aggregate and that there is often a differential spreading of the particles on the grid according to size. A better approach is to spray the suspension into a fine mist, collecting the droplets by holding the screen mounted on a microscope slide in the aerosol leaving the aerosol generator. The droplets should be smaller than 0.5 μm diameter. Since each droplet is smaller than the screen aperture, the residue from the evaporation of each droplet is easily traversed. Dautrebande,[42] and Cadle and Magill[43] have described aerosol generators that can be used for making such sprays. However, agglomeration remains a problem.

An approach that avoids agglomeration is to prepare such dilute suspensions of the particles that there is little likelihood that there will be more than one particle in each droplet of the spray. A highly volatile solvent can be used and the particles collected from the resulting aerosol by a technique such as thermal precipitation.

The powder can also be dispersed in the solvent in which the Parlodion or Formvar is dissolved prior to making the membrane, which is then prepared as described above.

Membrane filters are often used for collecting airborne particles to be measured by electron microscopy. Various methods have been used to mount on electron microscope grids the particles collected in this manner. One method involves placing small portions of the filters, with the particles on top, on the surface of a solvent for the filter material.[44] When all but thin membranes have been dissolved away, the membranes holding the particles are mounted on grids. This method is rather frustrating, because it is difficult to remove the membranes at just the right moment. Kalmus[45] has described a method for transferring the collected particles from a membrane filter to Formvar-coated grids. Small portions of the filter are placed on the grids with the particles against the Formvar membrane. The pieces of filter should be sufficiently small that they do not extend beyond the edges of the grids. The grids are next placed on strips of filter paper that have one end dipped in acetone and serve as wicks to bring the acetone in contact with the membrane filter material. The acetone dissolves the filter material, leaving the particles mounted on the Formvar membrane. Supports for the filter paper wicks can be made from rectangular strips of No. 100 stainless steel mesh bent in the form of an inverted U. The assembly is placed in a covered petri dish containing acetone. About 8 hours and several changes of acetone are required to remove all of the filter material. The screens bearing the particles are then placed on filter paper, dried overnight in a vacuum, and irradiated with ultraviolet light to increase the resistance of the Formvar films to electron bombardment.

Such techniques have been applied in the author's laboratory to polystyrene fiber filters with only partial success. It is difficult to remove the filter material without also removing an unacceptable amount of the particulate matter. As a result, the preparations tend to retain some polystyrene, and the images are not sharply focused and lack contrast. A better approach is to dissolve the filter in a suitable solvent such as methylene chloride, remove the particles by centrifuging, resuspend them in the solvent and centrifuge several times to remove the last traces of polymer, and finally allow a few drops of the suspension in the solvent to evaporate on a membrane-covered grid. Of course, the membrane must not be soluble in the solvent. The last step can be one of those described above for the mounting of powders.

Particles can sometimes be collected on electron microscope grids without the use of membranes. For example, the smokes of metal oxides produced by burning metals such as magnesium or zinc in air form very loose aggregates that are easily intercepted and retained by the wires of the grid. A more generally useful variation of this approach is to produce a network of very fine fibers over the grid and to use the grid plus fibers as an impactor by one of the methods described in the previous chapter. For example, air can be drawn through the array as though it were a filter. The smaller the fibers the more efficient will be the collection, and submicron-diameter fibers pulled with tweezers from filters are appropriate. A variation of this approach is to use individual fibers mounted on appropriate holders as impactors, again as described in the last chapter, and after sampling transfer the fibers with the particles to electron microscope grids.

Particles are often collected on smooth surfaces such as microscope slides or the anodes of electrostatic precipitators, and it later becomes desirable to measure the particles by electron microscopy. Sometimes the particles can be resuspended in some solvent, for example, using ultrasonic cleaning baths, and the dilute suspension concentrated by evaporation until the particle number concentration is relatively high. The slurry is then transferred to the membrane-covered screen by methods already discussed. However, if the particles have not all been redispersed, erroneous particle size distributions may be obtained. Another approach involves stripping the particles from the surface by a film which serves as the grid membrane. A few drops of a solution of Parlodion or Formvar are placed over the particles. The solution spreads into a film and is allowed to dry. The resulting membrane with the particles attached is then stripped off the surface and mounted on grids. Stripping the membrane can be rather difficult, but various methods developed for making surface replicas for study by electron microscopy can be applied. One method involves teasing the membrane from the surface using water to expedite the separation, which is started with a sharp knife or razor blade. The substrate such as a microscope slide on which the particles were collected is lowered into the water until the scraped-up edge of film, but not the upper surface, is wet. Sometimes the water alone is sufficient to free the membrane with its attached particles from the surface. In this case the membrane can be picked up as previously described. The membrane may have to be worked off, holding one edge with tweezers or a piece of Scotch tape. Hall[46] has described a method that obviates the need for immersing in water. The thoroughly dry film is moistened with water by breathing on it, and a number of grids are placed, convex side down, on the film. If necessary the grids can be fastened in place with a little cement applied to the edges. Small pieces of paper such as punched-out paper disks are placed over the girds and Scotch tape is pressed onto the preparation, one end at least extending beyond the edge of the film. This end is lifted and a drop of water is inserted between the film and the surface. The film is then lifted from the surface with the tape. The pieces of paper prevent the tape from sticking to the grids. The film around the grids is cut with a razor blade, and the grids are lifted away from the tape with forceps. Schaefer[47,48] has described a method for removing the film that does not require any water. The films are strengthened by applying an overlayer of gelatin, or the surface is preconditioned before coating with plastic. The gelatin method involves applying a 20% solution of gelatin dissolved in warm water to the surface coated with Formvar. The gelatin is dried with warm air, and as it dries it may split off, carrying the film with it. It may have to be worked off, using a scalpel or razor blade. The film is then laid on the surface of warm water, gelatin surface down, whereupon the gelatin dissolves, leaving the Formvar film floating on the water. This method has been used for making replicas of snow flakes.

2.3. Determination of Magnifications

Techniques for determining magnifications produced by an electron microscope have been discussed in detail in books on electron microscopy such as those by Fischer,[37] Wyckoff,[41] and Hall.[46] Replicas of diffraction gratings are often used as standards, and the separation of the spacings can be determined either spectroscopically or with an optical microscope. Since the intervals between the rulings of a grating are not precisely uniform, the magnification should be estimated from a large number of the rulings, a rather difficult procedure at high magnifications. This difficulty can be circumvented by calibrating the instrument at a low magnification, measuring the size of some small object, and using that object to calibrate the instrument at higher magnifications. The replicas can be made of a plastic such as Formvar by one of the methods described above, but such replicas are easily distorted by the electron beam. Distortion can be avoided with a positive replica made by depositing an evaporated carbon or silicon oxide film 100 to 200 Å thick on the plastic replica. The plastic is then dissolved in a solvent.[46] The magnification is usually determined for several projector lens settings, and a curve of setting versus magnification is constructed. This curve can be used as long as the objective pole piece, the length of the object holder, and the high voltage are not changed. Furthermore, when tap settings instead of actual currents are observed, slow changes in the circuits can produce undetected magnification changes. The magnification should probably be checked at least once a month.

Magnifications can also be determined with spheres of a plastic latex, such as Dow Latex 580G, which have a uniform diameter of about 2600 Å. Such spheres have been used both to calibrate instruments and as internal standards included in the field of view. The size of these spheres varies somewhat, depending on the method by which they are mounted, possibly as a result of the deposition of material on the spheres during mounting. Such spheres are not entirely uniform in size, but Williams and Backus[49] found that 70% of the diameters of 580G were in the range ±3% of 2590 Å. The Dow Chemical Co. has also produced mono-dispersed latexes with mean diameters as small as 880 Å.[50] Because of the variation in particle size, Hall[46] recommends that at least 100 images of the spheres be measured to determine the magnification.

2.4. Shadowing

Electron micrographs of particles are the projections of the cross sections of the particles onto a surface, and thus they have a two-dimensional appearance, unless stereo pairs are prepared. Small particles are at times difficult to observe unless the contrast is very high. Thickness measurements by focusing on the top and then on the bottom of a particle, which is practical in optical microscopy, is usually impossible in electron microscopy because of the large depth of field of the electron microscope. These disadvantages can be largely overcome by shadowing, which was briefly mentioned earlier. The technique consists of heating a small piece of an appropriate material such as a metal in an evacuated vessel, whereupon vapor of the heated substance travels in straight lines to the surface of the vessel and any interposed surface. The pressure must be sufficiently low that the mean free path of the residual air is less than the distance from the heated material to the walls of the vessel. When the vessel contains a mounted specimen, the vapor condenses on the specimen in a thin layer, and if the plane in which the particles lie forms an oblique angle with the direction of the stream of vapor, the deposition occurs on the portion of the particle surface not shaded from the vapor by the particle itself. Deposition also occurs on most of the surrounding membrane, but a metal-free region, the "shadow," lies behind each particle. Electron micrographs of shadowed particles are often printed as negatives rather than as positives so as to produce white particles against a dark background and even darker shadows. The result is a realistic three-dimensional appearance. The thickness of the shadowed particles can be calculated from the dimensions of the shadows and the angle between the specimen mount and the direction of flow of the vapor.

Metals are usually used for shadowing, and they should not produce grain comparable with the size of the particles. The deposited metal should not migrate over the surface and should produce adequate contrast with minimum thickness. Too thick a film obscures detail and can increase the apparent size of small particles. High atomic weight metals such as gold, uranium, platinum, and paladium are often used. Chromium has been found to be very satisfactory. A film of gold 5–10 Å thick is usually adequate. The correct thickness is usually determined by trial and error, and with a little experience the correct deposit thickness can be estimated from the color of the deposit on the microscope slides holding the mounts.

Specimens are sometimes shadowed from two directions to obtain even more information concerning particle shape than can be obtained from unidirectional shadowing.

The metal to be evaporated, in the form of fine wire or foil, is generally electrically heated with a tungsten filament. The vacuum is produced by a high-capacity diffusion pump backed by a mechanical pump. The vessel is usually a bell jar which rests on a metal plate. The author[1] has described very inexpensive equipment for shadowing that can be constructed in the laboratory.

2.5. Size Measurements

Measurements of particles with electron microscopes almost always involve measuring the images of the particles on plates, films, or prints by methods already described. The measurements are seldom made on the fluorescent screen of the microscope because of the low intensity of the illumination and the possibility of damage to the particles and the membrane on which they are mounted as a result of prolonged exposure to the electron beam.

Although the optical microscope cannot be used to measure particles much smaller than the wavelength of light, the electron microscope is not appropriate for the measurement of large particles because of the very small field of view even at the lowest magnifications for which most electron microscopes are designed. If the particle size distribution is very large, both optical and electron microscopy may be used to obtain the size distribution. The results obtained with the two instruments must be fitted together, and differences in the errors involved in using the two instruments introduce difficulties. The particle size ranges measured by the instruments should overlap, a great aid in combining the results.

One of the most thorough studies of the problems involved in combining the results of optical and electron microscopy was made by Cartwright,[53] using dust samples collected on membrane-coated grids by thermal precipitation. He emphasized the importance of collecting the electron microscope sample under precisely the same conditions as those for collecting the optical microscope sample. He also questioned whether electron microscope grids should ever be used in the thermal precipitator, since he found a preferential deposition of dust on the grid wires. Cartwright collected samples of dust by thermal precipitation on membrane-coated cover glasses. Measurements were made with an optical microscope, and the membranes were stripped from the cover glasses, mounted on grids, and transferred to an electron microscope. The cover glasses were coated with Formvar in a rather conventional manner by dipping them in a 0.7% solution of Formvar in ethylene chloride, but to ensure easy stripping, a prior layer of a hydrophilic substance was first formed on the cover glasses by dipping them in a 5% solution of Teepol. The density of the deposit produced by the thermal precipitator was selected for ease of measurement and counting with the optical microscope, and to avoid the possibility of obscuration of the field by overlapping particles. After measurements were made with the optical microscope, water was used to strip the membrane. The cover glass was held with tweezers at about 40° to the horizontal, and the lower edge was touched to the surface of distilled water for a few seconds. The cover glass was then slowly slid under the surface of the water, leaving the membrane floating on the surface. The membrane was removed from the water and transferred to a grid.

Cartwright found that the size distribution obtained by optical microscopy did not completely agree with that obtained by electron microscopy, and that compromises were necessary to combine the results. He concluded that the need for a compromise arose from the limitations of the optical microscope.

3. SCANNING ELECTRON MICROSCOPY

The scanning electron microscope has at least two major applications to fine-particle science and technology. The remarkable ability of this type of microscopy to reveal particle shapes and surface features was mentioned in Chapter 1 (see Figure 1-6). Scanning electron micrographs of grains of sand have been used to deduce the history of each grain examined.[54] Furthermore, most electron microprobes, described later, include low-resolving power scanning electron microscopes so that the appearance of each particle whose elemental composition is being determined can be observed.

Just as transmission electron microscopy is analogous to transmission optical microscopy, scanning electron microscopy has some of the features of flying-spot optical microscopy. The electrons are focused onto a very narrow portion of the field of view, and the latter is scanned with this "spot" in a raster, as in a television tube. Most, but not all, scanning electron microscopes operate with incident electron illumination. The incident electron beam knocks electrons from the surface and these are collected. The electrons emitted from each point on a particle surface are characteristic of that point, and serve to produce a magnified picture of the particle surface on a cathode ray tube. A block diagram of a scanning electron microscope is shown in Figure 3-32. Transmission scanning electron microscopes are more nearly analogous to flying spot optical microscopes in that the image is built up from electrons passing through the field of view. In general, the incident type of instrument is more useful for observing the surface characteristics of particles, and the remaining discussion deals with that type of microscope.

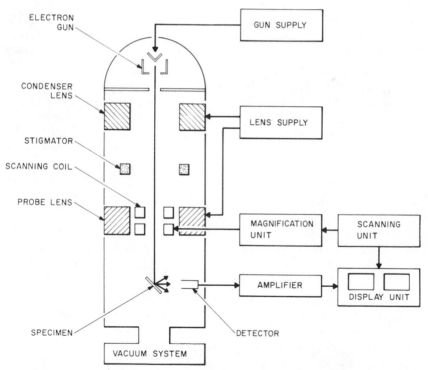

Figure 3-32.   Block diagram of a scanning electron microscope.

The beam of electrons is produced, as usual, with an electron gun, is focused to a small spot with a pair of condenser lenses, and is moved over the surface of the field of view with coils placed along its path. Usually the signal produced by the emitted electrons is used to modulate the brightness, as in Figure 1-6, but it can also be used to modulate the Y-deflection of the spot of an oscilloscope as shown in Plate XIV of Hawkes.[40] The latter approach produces a striking three-dimensional effect when applied to large surfaces, but some resolution may be lost.

The electron gun uses a heated tungsten filament, or a lanthanum hexaboride emitter, which produces greater brightness. For example, the Applied Research Laboratories' SEMA uses a self-biased triode gun which operates up to 50 kV. It is adjustable in the x and y directions, and the electron source is adjustable in the x, y, and z directions with respect to the Wehnelt cylinder. Electron emitters other than tungsten or lanthanum hexaboride can be inserted.

In addition to the two condenser lenses, a third, probe-forming, lens is used between the scanning coils and the specimens to shorten the length of the path of the electron beam. For a number of reasons, magnetic lenses are almost always used. The lenses are provided with apertures. The SEMA instrument is provided with externally selectable apertures. Those for the condenser lens are mounted in a thin sliding tube that is readily removed for cleaning and replacement.

The image contrast is determined by the variations in the numbers of electrons reaching the detector from various points on each particle. In addition to variations in surface topography, the elemental composition, electrical and magnetic fields on the particle surface, and crystallographic orientation can cause variations in the numbers of emitted electrons.

The electrons emitted by the particles produce a very small current, about 1 pA, which must be amplified. Conventional amplifiers are too noisy, and the amplification is generally achieved by the combined use of a scintillator and a photomultiplier. Electrons striking the scintillator produce light that is converted into an electrical signal by the photomultiplier and amplified in a conventional manner. The image produced by this signal is usually formed on each of two cathode ray tubes. One is arranged for visual observation and is coated with a fluorescent material with a relatively long persistence period, but which has rather large grains. The other tube is for preparing electron micrographs. The grain size is sufficiently small that the electron optics rather than grain size limit the resolution. The persistence time is short, but this is not important for photography.

Equipment for producing a vacuum is, of course, required, as for the transmission electron microscope, and a variety of stages can be used including tilting and hot stages.

Present scanning electron microscopes do not have the resolving power of transmission electron microscopes. The resolving power of the scanning instruments is limited by the size of the electron probe and in the highest-quality instruments is about 100 Å. The scanning microscope features of electron microprobes usually have an even lower resolving power, about 1000 Å.

Although secondary electrons are usually used to form the image, some of the primary electrons are scattered by the particles and can be used to form an image. The intensity of this scattering is markedly affected by the atomic number of the specimen; thus image formation by scattering electrons is especially appropriate when the microscope is being used to reveal variations in the elemental composition of the specimen.

The particles to be examined must not become charged by the impinging electrons, because irregular charging affects the path of both the incident and the secondary electrons. Thus the specimen must be conducting, and this is readily achieved by coating it with a thin layer of metal, as for shadowing. A thin layer of carbon is often satisfactory, and is especially appropriate if the elemental composition is to be determined by an electron microprobe.

Mounting specimens for observation with the scanning electron microscope is usually much simpler than mounting them for observations with the transmission electron microscope. This is a direct result of the fact that the former instrument scans the surface of the specimen, and thus the mounting need not be thin. For example, if the particles have been collected on a membrane filter, and are sufficiently large that all are on the top surface, portions of the filter can be coated with a conducting film and introduced directly into the microscope.

Many scanning electron microscopes can be equipped with an energy dispersive X-ray analyzer (EDXRA) to provide determinations of the elemental composition of the particles examined, as discussed in the next section.

Byers, et al.[55] described the application of a computerized method for size characterization of atmospheric aerosols by the scanning electron microscope. Samples were collected from the air of a number of central Pennsylvania communities by electrostatic precipitation. The particles were deposited either on aluminum foil with the bright side as the collector surface, or on aluminum-coated Mylar, which had a smoother surface. Imperfections on the foil were sometimes confused with particles by the computer, but this problem was nearly eliminated by placing the samples normal to the electron beam and orienting any striations in the direction of the detector. One method used for computer processing has been described by McMillan, et al.,[56] and another method investigated was that of Matson et al.[57] The former, simpler, method involved placing a star in a coordinate position corresponding to each data point whose intensity was above a given level. The fields of stars could represent the projected areas of individual particles lying on a substrate. The area of a given particle could be obtained by multiplying the number of points by the area corresponding to one point. Area calibration was accomplished by processing images of latex spheres of known diameter.

Figure 3-33 shows the ETEC Corporation "Autoscan" scanning electron microscope. The electron gun is the triode type, and the resolution is about 100 Å. The electronic components are in modules and are almost entirely solid-state. The system is evacuated with a water-cooled oil diffusion pump, eliminating the need for a liquid nitrogen cooled trap.

An innovation in scanning electron microscopy is represented by the Hitachi Perkin-Elmer HFS-2 Field Emission Scanning Electron Microscope. The electron emission by a thermal cathode gun is limited by the temperature before vaporization of the cathode occurs. In this microscope the usual gun is replaced with a field emission gun that extracts cathode electrons with an electric field in an ultrahigh vacuum. The resulting source is more than 1000 times as intense as a conventional thermal emitting source, and a much higher resolution can be achieved (30 Å for scanning incident electrons and 15 Å for scanning transmitted electrons).

The field emitter is the tip of a tungsten needle produced by polishing a single crystal wire in an electric field. The tip radius is about 1000 Å.

4. THE ELECTRON MICROPROBE

Although the electron microprobe is primarily an analytical instrument and thus somewhat outside the scope of this book, it can be used to obtain, automatically, particle size distributions of those particles containing a large percentage of some element in a mixture in which many of the particles contain little or none of that element. The electron microprobe is a very close relative of the scanning electron microscope. A very narrow electron beam from an electron gun is focused on a par-

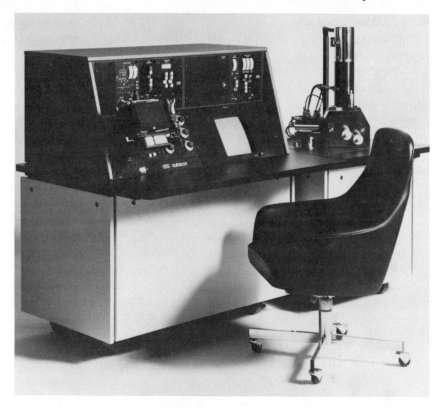

Figure 3-33. The ETEC Corporation "autoscan" scanning electron microscope.

ticle or surface to be analyzed, whereupon the surface fluoresces, emitting X-rays whose wavelengths are determined with a sensitive X-ray spectrometer. Each element emits X-rays having wave lengths characteristic of that element. Several X-ray spectrometers are usually used to analyze the wavelength range of interest. Crystals usually serve as diffraction gratings, and are rotated to focus various wavelengths on the detector.

The X-rays produced by an electron beam are a consequence of the removal of inner electrons from the elements in the particles. This removal is followed by the shift of electrons from outer shells into K, L, or M inner shells. Each such shift produces a photon of X-radiation characteristic of the element in which the shift occurs. The wavelength range is very narrow, so the spectrum from any given element consists of a few sharp lines. Furthermore, the spectrum is little affected by the compounds, alloys, or minerals in which the element occurs. Unfortunately, continuous wavelength (white) radiation is also emitted, produced by the deflection and deceleration of electrons that have penetrated the atoms until they are close to the nucleus. This background radiation is called bremsstrahlung or braking radiation and can markedly affect the sensitivity of electron microprobes. The amount of bremsstrahlung increases with increasing atomic number; mounting particles on heavy metal surfaces, such as platinum microscope slides, for examination by electron microprobe analyzers should be avoided.

The electron microprobe can be used in at least two modes. One involves focusing the beam of electrons on a single particle or position on a surface and determining the X-ray spectrum for a large number of elements. This often involves changing the spectrometer crystals once or twice to include the entire range of wavelengths. The other mode involves setting the spectrometer on a single wavelength, corresponding to a specific element, and scanning the entire field with the electron probe as is done with scanning electron microscopy. The output from the spectrometer detector produces an image on a cathode ray tube.

Most electron microprobes can also be used as low resolution electron microscopes (limit of resolution about 1000 Å) and can present an image of the field of view, produced by the secondary electrons, on a cathode ray tube adjacent to the tube producing the image in terms of a particular element. The images can, of course, be photographed to provide a permanent record. Figure 3-34 shows the type of record obtained. Figure 3-34a is the electron microprobe image of a copper particle produced by Cu $K\alpha$ X-rays, and 3-34b is the scanning electron micrograph of the same particle taken at the same time.

Electron microprobes include a light microscope as an integral part of the system for selecting fields for study and obtaining a preliminary view of the specimen at a relatively low magnification. Because of space limitations in the body of an electron microprobe, the light microscope must be designed to compromise between excellent light microscopy and the restrictions imposed by the microprobe configuration, which of course takes precedence. The light microscopes of some microprobes have mirrors as objectives. Central openings in the mirrors allow the electron beam to pass. Other instruments use more conventional objectives which can be moved out of the path of the electron beam. Usually only one or two magnifications are provided.

Although curved crystals are commonly used as diffraction gratings to resolve the spectra, ruled gratings are sometimes used. They have the

Figure 3-34. Electron microprobe image with Cu Kα x-rays (A) and scanning electron micrograph (B) of a copper particle obtained with the MAC 400 electron microprobe. Distance across each micrograph = 90 μm.

advantage that a single grating can resolve the entire wavelength range of interest, but they have a low scanning speed.

The newer electron microprobes are also equipped with energy dispersion X-ray analyzers (EDXRA), mentioned in the preceding section in connection with scanning electron microscopes. The sensing elements are photosensitive solid-state diodes that produce an electrical impulse for every X-ray photon that impinges upon them. The intensity of the pulse is related to the photon energy, hν, and the pulses are stored in multichannel analyzers from which the data can be retrieved and displayed on oscilloscopes. The advantages of the EDXRA are the high rates at which they detect the elements and of course the fact that crystal gratings do not have to be changed. Furthermore, because of the large angle of X-ray collection (essentially, a large angular aperture), the X-ray intensity is orders of magnitude higher for the EDXRA than for crystal spectrometers. However, McCrone and Delly[18] point out that the resolution of even the best of the solid-state detectors is no better than 150–180 eV as compared with 2–10 eV for crystal spectrometers. As a result of the poor resolution by the EDXRA, interference may occur between elements with adjoining lines, and the presence of such elements may have to be verified with a crystal detector.

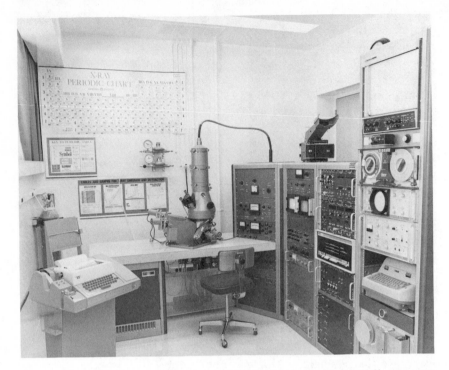

Figure 3-35.   MAC 400 electron microprobe combined with a small computer at the National Center for Atmospheric Research.

The electron microprobe can be automated to yield particle size distributions for the entire particulate matter and also for that portion of the particulate matter containing a large percentage of some selected element. At the National Center for Atmospheric Research, the electrical signals from an MAC electron microprobe are fed into a small computer which transfers them into a form on magnetic tape such that they can be directly analyzed by the large Control Data 6600 or 7600 computers. The final output is a size distribution, usually based on a definition of particle diameter as the diameter of the circle having an area equal to the projected area of the particle. Figure 3-35 is a photograph of the system. The same computer system is also used to automatically determine size distributions with a light microscope. A television camera is mounted at the eye point of a conventional microscope and the output displayed on a cathode ray tube by closed-circuit television. At the same time, the signal is fed into the small computer and analyzed in the same manner as the signals from the electron microprobe.

McCrone Associates have developed a simplified electron microprobe that they call the Miniprobe.[18] It was developed to overcome the high cost of electron microprobes and the requirement of highly skilled personnel to operate them. It is claimed that it can analyze 90% or more of the particles encountered routinely.

The surface on which the particles are mounted should be smooth and electrically conducting in addition to having a low atomic number. Common substrates are beryllium and carbon. The carbon can be evaporated onto a surface such as a microscope slide to produce the desired substrate. An electron microscope grid covered with a carbon film can be used for both electron microprobe analysis and transmission electron microscopy. A thin film of carbon evaporated over the particles can be used to render the particles conducting, although for very small particles this may not be necessary.

5. MISCELLANEOUS METHODS

5.1. Determination of Collected Mass

Determining the mass concentration of airborne particulate matter is usually a relatively simple undertaking involving a simple weighing, but even so, certain precautions should be taken. If the particulate material is hygroscopic or deliquescent, a decision must be made as to whether the result is to be obtained for the anhydrous material, for some specific relative humidity (or series of relative humidities), or for the average relative humidity of the air at the time the sample was collected. A similar decision must be made with regard to temperature. Then the sample must be weighed under the conditions of interest, allowing sufficient time for equilibrium to be reached. Equilibrium may be approached only after a matter of hours, and the attainment of equilibrium can be verified by repeated weighing. Another problem arises when the particulate material contains somewhat volatile or easily oxidized organic matter. When collecting and weighing such material, for example from city air, the sampling time should be no longer than necessary to collect sufficient material to weigh (and possibly to analyze). The sample should be weighed as soon as possible following collection.

Glass fiber filters are commonly used for determining the mass concentrations of airborne particles, and are especially appropriate for this purpose because they do not adsorb appreciable moisture from the air. The usual approach is, of course, to weigh the filter before and after sampling. Many other types of filters can be used. Fiber filters are especially useful because they have a low resistance to air flow, making very high sampling rates practical, and because they have a very large collection efficiency even for very small particles. The moisture-adsorbing tendency (the adsorption isotherm) of a filter should be known or determined before its use for obtaining mass concentrations of airborne particles. If this tendency is large, the filter may have to be weighed at the same temperature and relative humidity before and after sampling.

Gases in addition to water may also be adsorbed or absorbed, and the amounts may be comparable to those of the particles. If the collection efficiency for the gas is small, two filters can be used in series and the second used as a blank. However, a better approach is to use filters that have been pretested to ensure that their adsorption capacity is small.

The material collected by sedimentation in "dust jars" is usually weighed and the result converted to a mass concentration in the air. The dust jars themselves can be weighed before and after sampling, but usually the particles are removed and weighed, for example, by washing the particles from the jars and evaporating the wash water.

Particles collected in any manner can, of course, in principle be weighed. For example, samples collected by thermal precipitation or impaction can be weighed with a microbalance. Mass concentrations for such samples are often calculated from a known or assumed particle material density, the particle size distribution, and the particle number concentration. The latter method often produces quite inaccurate results, especially since a few large particles are often responsible for much of the mass of the sample. Of course, even direct weighing will produce erroneous mass concentrations unless the sample is representative of the aerosol sampled, but the error is often compounded by the additional sampling involved in selecting fields of view for particle measurement and counting by microscopy.

Mass concentrations can often be determined by chemical analytical means. For example, if aerosol particles consist entirely of sodium sulfate, they can be collected by some means, dissolved in water, and the solution analyzed for sulfate ion. If an aerosol contains a mixture of chemical species, the mass associated with one or more of these species can often be determined by some type of chemical (or physical) analysis. However, the methodology of such analytical procedures is beyond the scope of this book.

5.2. Light-scattering Techniques

Particles collected on microscope slides or other transparent substrates are sometimes investigated by extinction and light-scattering methods. A beam of light normal to the slide is passed through the collected particles and the decrease in light intensity produced by the particles (the extinction) or the intensity of the scattered light is measured. The theory and the techniques employed are essentially the same as for measurements of extinction and light scattering directly on the suspended particles. Chapter 4 is devoted to this subject.

There are numerous difficulties to measuring particles collected on microscope slides by light scattering or light extinction. The percentage extinction is usually small, the deposit is often irregular, and dirt on the slide can cause large errors. A technique that overcomes some of these difficulties involves depositing a thin film of a metal such as aluminum on the sample by the shadowing technique, but with the beam of vapor normal to the collection surface, followed by removal of the particles. The light transmitted by the holes produced is theoretically equal to the light incident upon them for values of $2\pi r/\lambda$ down to about two for circular holes, and experimental evidence suggests that this relationship holds even for irregular holes down to at least $2\pi r/\lambda = 6$. The angular distribution of the transmitted light approximates that calculated from Fraunhofer diffraction theory, and methods described in Chapter 4 to measure extinction can be used to obtain mean sizes of the holes and thus of the particles.

When particles are collected on an intermittently moving tape, a series of dark spots is often obtained, each spot being produced by the sample. The amount of darkening produced is indirectly related to the particle loading of the atmosphere, although it is also markedly affected by the composition of the particles and by the size distribution. Obviously, for the same mass concentration of particulate material, an atmosphere containing soot will produce darker spots than one containing wind-blown dust. The contamination level can be crudely estimated by visually comparing the spots with standards, but a more precise approach is to determine the amounts of light transmitted through or reflected from each spot. Such quantitative measurements are usually employed.

The use of the darkening of a filter by polluted air to indicate the degree of pollution has long been a popular method. Linsky[58] mentions a recent reprinting of the 1915 Chicago Air Pollution Summary Report that shows a 1912 mobile air sampling laboratory truck using the color of filtrate as an air pollution index.

Hemeon et al.[59] developed the concept of "coefficient of haze" for the evaluation and interpretation of filter paper samples. It is based on Beer's Law:

$$\log \frac{I_0}{I} = kQ \equiv \text{optical density} \qquad (3\text{-}6)$$

where k is a constant, Q is the quantity of particulate material in suspension, I_0 is the intensity of the light incident on a suspension, and I is the intensity of the light transmitted by the suspension. To determine whether this relationship applies to particulate material collected on filter paper, outdoor air was aspirated simultaneously through a number of paper filters (Whatman No. 4) at varying flow rates for a convenient time period. Thus the quantity of collected particulate material was proportional to the total amount of air drawn through the filter. Then with a very simple, homemade, light transmission apparatus, and defining I_0 as the intensity of the light passing through a clean filter, optical densities were determined and plotted against the quantity of air filtered. Straight lines were obtained for small optical densities, showing that the results conformed to Beer's law. No important divergence from the straight line relationship occurred for optical densities below about 0.03 (light transmission not less than about 50%).

Hemeon et al. defined a coefficient of haze or "Coh" unit as that quantity of light-scattering solids producing an optical density of 0.01 when measured by light transmission. Thus an optical density of 0.301 is 30.1 Coh units. Furthermore, they found it convenient to reduce all measurements to Coh units per thousand linear feet of air. The optical density is simply divided by the number of thousands of feet sampled. This unit is being used by many laboratories to report the results obtained with tape samplers.

These authors also made some measurements by light reflectance and plotted optical density by light transmission against "optical density by reflectance." Straight lines were obtained, but the data indicated that transmission measurements were more sensitive for light deposits.

Gruber and Alpaugh[60] preferred to use reflectance measurements for evaluating the spots on tape samplers, and proposed the unit that they christened the "RUDS" (Reflectance Units of Dirt Shade). Dirt shade was defined by the equation

$$\text{Dirt shade} = \log \frac{\% \text{ reflectance clean filter paper}}{\% \text{ reflectance soiled filter paper}} \times 100. \qquad (3\text{-}7)$$

Thus absolutely clean air produces a value of zero. The authors had never observed a dirt shade exceeding 100 RUDS. They choose to use

RUDS per 10,000 linear feet instead of per 1000 linear feet and report the values in whole numbers. In practice, reflectance meters can be set at 100 when receiving light from the clean filter. The reflectance indicated for the soiled filter paper is then the percentage reflectance for that filter. Like the Coh, the RUDS is often used for reporting results obtained with tape samplers.

Numerous studies have been made of the relationships between values of the coefficient of haze (soiling index) and other indicators of air pollution intensity. Burt[61] attempted to establish a relationship between visibility and the Coh per thousand feet, using data obtained in St. Louis and Cincinnati. Some of the results obtained for St. Louis are shown in Figure 3-36. Obviously, although a positive correlation existed between the Coh and visibility, the correlation coefficient was low. Furthermore, the visibility was found to be a function of the relative humidity, whereas the Coh was not. Burt concluded that although the visibility-soiling index-humidity correlations were significantly different from zero, they were not sufficiently high to permit satisfactory estimates of the Coh from measurements of visibility and humidity, even in a single city where the type of pollution is more or less fixed.

Hale and Waggoner[62] determined the variations of Coh values from Type D AISI (American Iron and Steel Institute) tape samplers, prod-

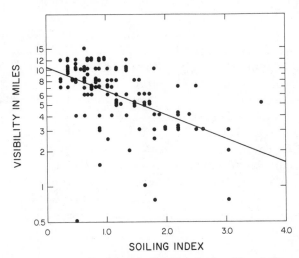

Figure 3-36. Log-law relationship of visibility and soiling index at 60% relative humidity (1200 CST, winter, St. Louis).[61] Courtesy of the American Industrial Hygiene Association.

ucts of the Research Appliance Co. Twelve adjacent samplers were operated in a location where expected values would be generally less than
3.0, in order to estimate gross errors in sampling and readout under
suburban sampling conditions. The machines used were relatively new
(received within 18 months) and had been relatively trouble-free. The
densitometers were of the Research Appliance Company Model 157 type.
Standard deviations for a series of single observations made by each of
four "readers" varied among the readers from 0.195 to 0.247, and the
standard deviations for averages of twelve observations varied from
0.048 to 0.092 among the four readers. Hale and Waggoner concluded:
(1) the variability of the Coh unit data may vary from reader to reader,
(2) the largest source of variation is the variation in the light transmission through the paper, and (3) highly competent readers should be
selected. Double reading of tapes and the training of readers was
recommended.

Saucier and Sansone[63] experimentally established relationships between transmittance (Coh units) and reflectance (RUDS units) measurements for aerosols of constant optical properties and for airborne particles in Pittsburgh, Pennsylvania. The study was undertaken because
some air pollution control bureaus evaluate tape samples by transmittance, and others evaluate them by reflectance. Furthermore, in 1970 the
National Air Pollution Control Administration recommended the use
of transmittance measurements for the Warning-Alert-Emergency sections of local ordinances, whereas many of these ordinances are written
in terms of reflectance.

Particulate material was collected with calibrated AISI samplers. Suspensions of dusts were dispersed in a 9-m^3 chamber by means of a com-
pressed-air driven dust feeder. The dusts were coal, limestone, flyash, and
a 1:2 by weight coal-limestone mixture. Measurements were also made
at three sites in Allegheny County, Pennsylvania: one suburban site and
two urban sites in residential areas.

The results were plotted as percentage reflectance versus percentage
transmittance. The plots were essentially straight lines, but the slopes
varied markedly for the different types of dust. The correlation coefficients ranged from 0.87 to 0.99. The slopes for the three field locations
were nearly identical, and the relationships between the transmittance
and reflectance measurements could be expressed as

$$1 \text{ Coh} \simeq 0.5 + 0.1 \text{ RUDS} \tag{3-8}$$

The correlation coefficients for the urban sites ranged from 0.92 to 0.99.

Linsky,[58] in a discussion of the Saucier-Sansone paper, strongly criticized such tape measurements as a basis for air pollution control, point-

ing out that they cannot take into account the color and light-scattering properties of the particulate pollutants in addition to their mass concentrations.

Of course, methods other than light-scattering or transmittance can be used to evaluate the tapes. For example, the pressure drop across the tape can be determined by measuring the pressure in the vacuum line downstream of the filter. The pressure drop will, of course, increase with increasing particle loading. The change in the transmittance of the tape to beta radiation as the particle loading increases has also been employed. A source of beta radiation is placed on one side of the tape, and a Geiger counter or similar device is placed on the other side.

5.3.   Methods for Large Amounts of Particulate Matter

A number of methods for determining particle size distributions are available for relatively large samples of particulate matter, usually in powder form. Such methods are the ones most often used for determining the size distributions of powders, but unless greatly modified they are seldom applicable to airborne particles. Exceptions to this generality can occur when large amounts of particulate material are collected from heavily loaded atmospheres such as flue and stack gases. Examples of methods mainly applicable to powders are sieve analysis, elutriation, liquid sedimentation, and surface area measurements. Such methods are described in detail in numerous books devoted to particle size determination in general, and because of their limited applicability to airborne particles will not be further considered here.

REFERENCES

1. Cadle, R. D., *Particle Size Determination,* Interscience, New York, 1955.
2. Allen, T., *Particle Size Measurement,* Chapman & Hall, London, 1968.
3. Irani, R. R., and C. F. Callis, *Particle Size: Measurement, Interpretation, and Application,* Wiley, New York, 1963.
4. Orr, C., Jr., and J. M. Dalla Valle, *Fine Particle Measurement: Size, Surface and Pore Volume,* MacMillan, New York, 1959.
5. Chamot, E. M., and C. W. Mason, *Handbook of Chemical Microscopy,* V. 1, 3rd ed., Wiley, New York, 1958.
6. Flügge, J., *Zeiss Information,* 15 (No. 64), 69 (1967).
7. Allen, R. M., *The Microscope,* Van Nostrand, New York, 1940.
8. Clark, G. L., ed., *The Encyclopedia of Microscopy,* Reinhold, New York, 1961.

9. Whytlaw-Gray, R., and H. S. Patterson, *Smoke,* Edward Arnold & Co., London, 1932.

10. Bennett, A. H., J. Jupnik, H. Osterberg, and O. W. Richards, *Phase Microscopy. Principles and Applications,* John Wiley, New York, 1951.

11. Richards, O. W., *Science,* **120,** 631 (1954).

12. Needham, G. H., *The Microscope,* Charles C Thomas, Publisher, Springfield, Ill., 1968.

13. Francon,M., *Progress in Microscopy,* Row, Peterson and Co., Evanston, Ill., 1961.

14. Grey, D. S., *J. Opt. Soc. Am.,* **39,** 719, 723 (1949).

15. Grey, D. S., *J. Opt. Soc. Am.,* **40,** 283 (1950).

16. Schaeffer, H. F., *Microscopy for Chemists,* Dover, New York, 1966.

17. Stoiber, R. E., and S. A. Morse, *Microscopic Identification of Crystals,* The Ronald Press, New York, 1972.

18. McCrone, W. C., and J. G. Delly, *The Particle Atlas,* 2nd ed., Ann Arbor Science Publishers, Inc., Ann Arbor, Mich., 1973.

19. Crossmon, G. C., *Anal. Chem.,* **20,** 976 (1948).

20. Grabar, D. G., *J. Air Poll. Control Assoc.,* **12,** 560 (1962).

21. Cherkasov, Yu. A., *Intern. Geol. Rev.,* **2,** 218 (1960).

22. Patterson, H. S., and W. Cawood, *Trans. Faraday Soc.,* **32,** 1084 (1936).

23. Fairs, G. L., *Roy. Micros. Soc. J.,* **31,** 209 (1951); *Chem. Ind.,* **62,** 374 (1943).

24. May, K. R., *J. Sci. Inst.,* **22,** 187 (1945).

25. Watson, H. H., *Brit. J. Indus. Med.,* **9,** 80 (1952).

26. Hamilton, R. J., J. F. Hodsworth, and W. H. Walton, *Brit. J. Appl. Phys. Suppl.,* **3** (1952).

27. Watson, H. H., and D. F. Mulford, *Brit. J. Appl. Phys.,* Suppl. 3 (1954).

28. Bovey, D., *J. Sci. Inst.,* **39,** 405 (1962).

29. Green, H., *Ind. Eng. Chem.,* **16,** 677 (1924).

30. First, M. W., *Arch. Ind. Hyg. Occ. Med.,* **7,** 58 (1953).

31. The Institute of Physics, The Physics of Particle Size Analysis, *Brit. J. Appl. Phys.,* Suppl. 3, London, 1954.

32. Roberts, F., and J. Z. Young, *Proc. Inst. Elec. Engrs.,* **99,** Part IIIA, 747 (1952).

33. Ames, D. P., R. R. Irani, and C. F. Callis, *J. Phys. Chem.,* **63,** 531 (1959).

34. Braham, R. R., B. K. Seely, and W. D. Crozier, *Tr. Am. Geophys. U.,* **33,** 825 (1952).

35. Holden, F. R., F. W. Dresch, and R. D. Cadle, *A.M.A. Arch. Ind. Hyg. Occ. Med.,* **9,** 291 (1954).

36. Robinson, E., J. A. MacLeod, and C. E. Lapple, *J. Meteorol.,* **16,** 63 (1959).

37. Fischer, R. B., *Applied Electron Microscopy,* Indiana University Press, Bloomington, 1954.

38. Kynaston, D., and J. Mulvey, in *Electron Microscopy. Proceedings of the Fifth International Congress for Electron Microscopy,* Academic Press, New York 1962.

39. Wilska, A. P., in *Electron Microscopy. Proceedings of the Fifth International Congress for Electron Microscopy,* Academic Press, New York, 1962.

40. Hawkes, P. W., *Electron Optics and Electron Microscopy,* Taylor and Francis, Ltd., London, 1972.

41. Wyckoff, R. W. G., *Electron Microscopy Technique and Applications.* Interscience, New York, 1949.

42. Dautrebande, L., *Arch. Intern. Pharmadynamie,* **84,** 1 (1950).

43. Cadle, R. D., and P. L. Magill, *Ind. Eng. Chem.,* **43,** 1331 (1951).

44. First, M. W., and L. Silverman, *Arch. Ind. Hyg. Occ. Med.,* **7,** 1 (1953).

45. Kalmus, E. H., *J. Appl. Phys.,* **25,** 87 (1954).

46. Hall, C. E., *Introduction to Electron Microscopy,* 2nd ed., McGraw Hill, New York, 1966.

47. Schaefer, V. J., *Phys. Rev.,* **62,** 495 (1942).

48. Schaefer, V. J., *Science,* **97,** 188 (1943).

49. Williams, R. C., and R. C. Backus, *J. Am. Chem. Soc.,* **71,** 4052 (1949).

50. Bradford, E. B., and J. W. Vanderhoff, *J. Appl. Phys.,* **26,** 864 (1955).

51. Fisher, C., *The Microscope,* **19,** 1 (1971).

52. Jesse, A., *The Microscope,* **19,** 21 (1971).

53. Cartwright, J., *Brit. J. Appl. Phys. Suppl.,* **3,** 109 (1954).

54. Krinsley, D. H., and I. J. Smalley, *Am. Scientist,* **60,** 286 (1972); *Science,* **180,** 1277 (1973).

55. Byers, R. L., J. W. Davis, E. W. White, and R. E. McMillan, *Environ. Sci. and Tech.,* **5,** 517 (1971).

56. McMillan, R. E., G. G. Johnson, Jr., and E. W. White, Proceedings of the 2nd Annual Electron Microscope Symposium, IITRI, 439 (1969).

57. Matson, W. L., E. W. White, and G. G. Johnson, Jr., 1970 Symposium on Computational Photogrammetry, Washington, D.C., 1970.

58. Linsky, B., *Atmos. Environ.,* **6,** 871 (1972).

59. Hemeon, W. L. C., G. F. Haines, Jr., and H. M. Ide, *Air Repair (J. Air Poll. Control Assoc.),* **3,** 22 (1953).

60. Gruber, C. W., and E. L. Alpaugh, *Air Repair (J. Air Poll. Control Assoc.),* **4,** 143 (1954).

61. Burt, E. W., *Am. Ind. Hyg. Assoc.,* **22,** 102 (1961).

62. Hale, W. E., and N. E. Waggoner, *J. Air Poll. Control Assoc.,* **12,** 322 (1962).

63. Saucier, J.-Y., and E. B. Sansone, *Atmos. Environ.,* **6,** 37 (1972).

4

OPTICAL MEASUREMENTS OF AEROSOLS

The optical methods for measuring airborne particles described in the previous chapter required collection of the particles. In this chapter methods that do not require prior particle collection are described and discussed. Most, but by no means all, of the methods are based on the measurement of that fraction of an incident beam of light falling on an aerosol which is transmitted, or on the properties of light scattered by an aerosol or by individual particles. Some methods involve forming images of the particles, and thus are a type of optical microscopy, but most methods do not involve image formation. The output of many of the sensors employed is an electrical impulse, a fact which greatly simplifies automation.

1. LIGHT SCATTERING

1.1. Theory

Observable light is usually scattered, either by particles or by surface irregularities, and particles that remain airborne for appreciable times, are for the most part, very effective for scattering visible radiation. Furthermore, many particles absorb light, and the extent of this absorption is quantitatively indicated by the imaginary part of the complex refractive index of the particulate material. The complex index of refraction can be expressed as $n(1 - ik)$. The imaginary part, nk, is called the damping factor, and k is called the index of absorption or the index of attenuation.[1] The scattering and absorption of light by particles in the atmosphere are believed to play an important role in the earth's radiation balance and hence in climate.

General equations have been derived for predicting the various features of light scattered by particles of all sizes, but for convenience scattering theory is usually considered in terms of size range relative to the wavelength or wavelength range of the incident light.

The most general equations were derived by Mie[2] and are based on theory developed by Maxwell. Thus these equations are usually termed the Mie equations and the corresponding theory Mie theory. However, Kerker[1] points out that Mie's paper was preceded by independent solutions to the problem of light scattering, starting with that by Lorenz.[3]

Mie theory is quite complicated for particles of size about equal to that of the wavelength of the illuminating radiation, but becomes much simpler for particles much smaller or much larger than the wavelength. When the particles are much smaller, the equations assume a form originally derived by Lord Rayleigh, and when they are much larger the particles scatter on the basis of Kirchhoff diffraction, external reflection, and transmission with refraction.

Scattering theory as usually applied is based on the assumption that the scattering from a suspension can be considered to be the sum of the scattering by the individual particles. When the aerosol volume is small and the particle number concentration not especially large, this assumption is warranted. Usually only such "single scattering" need be considered when evaluating the results of particle measurements obtained with instruments based on light scattering. An example of a situation for which multiple scattering may be important is the interaction of sunlight with large clouds such as stratus.

1.1.1. Rayleigh Scattering

The theory for the scattering of light by particles that are small relative to the wavelength of light was first described by Lord Rayleigh[4] to account for the blue color of the sky. It is usually written in the form

$$s = 24\pi^3 \left(\frac{m^2 - 1}{m^2 + 2}\right)^2 \frac{V^2}{\lambda^4} \tag{4-1}$$

where s is the total amount of light of wavelength λ scattered by a sphere of volume V and refractive index m (relative to air) per unit intensity of light. This equation is derived for spherical, nonabsorbing particles of radius less than about 0.1λ. The equation also applies to very small particles such as molecules, which are not spherical. Thus blue light is scattered much more than light of longer wavelengths by air molecules and the sky appears blue.

When the incident light is unpolarized, the intensity scattered at an angle γ to the incident beam is

$$I = \frac{9\pi^2}{2R^2}\left(\frac{m^2-1}{m^2+2}\right)^2 \frac{V^2}{\lambda^4}(1+\cos^2\gamma) \tag{4-2}$$

where R is the distance from the particle to the point of observation and must be large relative to the particle size (Figure 4-1). Note that the right side of equation (4-2) is actually the sum of two terms, the multiplier of one being unity and of the other $\cos^2\gamma$. The former determines the intensity of the vertical (i_1) and the latter of the horizontal (i_2) polarized component of the scattered light, the terms vertical and horizontal referring to the plane determined by the point of observation and the direction of the beam of light.

An efficiency factor for scattering (scattering coefficient), Q_{scat}, can be defined as the total light scattered by a particle in all directions divided by the cross sectional area, $s/\pi r^2$:

$$Q_{scat} \text{ (Rayleigh scattering)} = \frac{128\pi^4 r^4}{3\lambda^4}\left(\frac{m^2-1}{m^2+2}\right)^2 \tag{4-3}$$

or using the dimensionless parameter

$$\alpha = 2\pi r/\lambda \tag{4-4}$$

$$Q_{scat} \text{ (Rayleigh scattering)} = \frac{8}{3}\alpha^4\left(\frac{m^2-1}{m^2+2}\right)^2 \tag{4-5}$$

For Rayleigh scattering, the angular distribution of the intensity of the scattered light is symmetrical about a plane normal to the illuminating beam. The distribution for the two polarized components of the scattered light, i_1 and i_2, is shown in Figure 4-2.

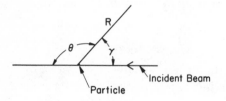

Figure 4-1.   Definition of R, γ, and θ.

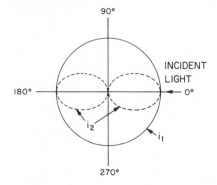

Figure 4-2. Angular distribution of the intensity of light scattered in the "Rayleigh region."

Light scattering results from the oscillating field of the radiation inducing an electric moment in the particles. This moment acts as a linear oscillator that radiates light in all directions except the direction of vibration. Rayleigh's equation was derived for optically isotropic particles (defined in Chapter 3). When the particles are anisotropic, the direction of the electric field of the incident radiation will not necessarily coincide with the direction of the induced moment, in which case the light scattered at 90° to the incident light will not be completely vertically polarized but will have a weak horizontal component.

1.1.2. Mie Theory

Light scattered by a particle comparable in size or large relative to the incident light is the resultant of light waves originating from various regions of the particle. The phase and intensity of these light waves are related in a complicated manner, and the nature of the light scattered from such particles is predicted by Mie theory. When the particles are very small, the Mie equation simplifies to yield the Rayleigh equation. The equation is actually a series of spherical harmonics terms, the coefficients of which are m and α. Except for particles for which the Rayleigh equation applies, the summation of the series can be very tedious, although modern computers have greatly simplified the problem. The derivation of the Mie equation and methods of evaluating it have been discussed in detail by van de Hulst[5] and by Kerker.[1] Extensive bibliographies of numerical computations based on Mie theory are provided by these authors and also by Hodkinson.[6]

Scattering coefficients (Q_{scat}) can, of course, be calculated for Mie scattering as well as for Rayleigh scattering. The complexity of the variation of the scattering coefficient with α and with the refractive in-

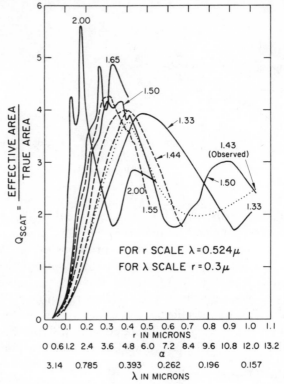

Figure 4-3. Scattering coefficients for spherical particles as a function of λ, r, and the refractive index as shown on each curve.[7]

dex for single isotropic spheres is demonstrated by Figure 4-3, prepared by Sinclair.[7] The approach to Rayleigh scattering as α becomes small is apparent, and so is the decrease in the value of α for maximum scattering with increasing refractive index. The angular distribution of the intensity of the scattered light also varies with α, and above the Rayleigh region as α increases the forward scattering increases relative to the backward scattering until for very large values of α the backward scattering is negligible.

The validity of Mie theory has often been demonstrated experimentally. An example is the work of Tabibian et al.[8,9] The dotted curve in Figure 4-3 was experimentally determined for comparison with the theoretical curves.[7] Gucker and Rowell[10] determined the angular variation of light scattered by single dioctyl phthalate aerosol droplets and compared the results with theoretical calculations. The particles were charged, suspended in an electrostatic field, and illuminated with monochromatic light. The light scattered was measured over the range 40 to 140° from the direction of illumination (θ, Figure 4-1). This work was continued by Gucker and Egan,[11] and some of their results are shown in Figures 4-4 to 4-6. Even more precise comparisons were made by Phillips et al.[12] The particles were polystyrene latex spheres that were charged and suspended in an electrostatic field. A radial electrostatic field kept the particles from drifting horizontally. The scattered light was measured as a function of angle with a Differential II light-scattering photometer (Science Spectrum, Inc.). Theoretical analysis of the measured intensities yielded values of the refractive index and size with relative errors of less than one percent.

Few particles are transparent isotropic spheres, and absorbing particles have a complex refractive index. The Mie equation has now been solved for many different types of particles. Van de Hulst[5] has reviewed the calculations for spheres of a number of metals including iron, nickel, zinc, copper, and carbon in air; and gold, silver, and mercury in water. Kerker[1] has provided a very detailed discussion of light scattering by infinite cylinders, including the effects of orientation, diameter, and refractive index. Many of the calculations were made by Kerker and his students. When illuminated by a beam of light, an infinite cylinder,

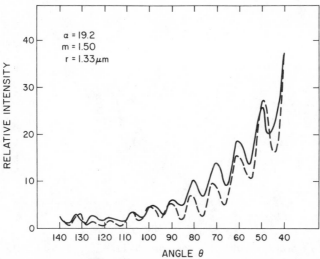

Figure 4-4. Scattering diagram for particle of 1.33 μm radius illuminated by unpolarized light. Solid line, experimental; dashed line, theoretical.[11]

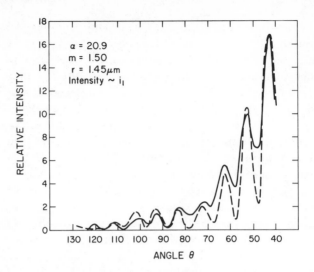

Figure 4-5. Scattering diagram for particle of 1.45 μm radius illuminated by plane polarized light with electric vector perpendicular to plane of observation. Solid line, experimental; dashed line, theoretical for i_1.[11]

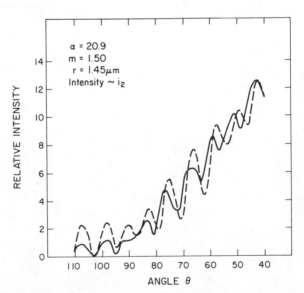

Figure 4-6. Scattering diagram for particle of 1.45 μm radius illuminated by plane polarized light with electric vector parallel to plane of observation. Solid line, experimental; dashed line, theoretical for i_2.[11]

regardless of orientation, does not scatter radiation in all directions as does a sphere. The radiation scattered by a small portion of the cylinder surfaces (element) lies in the surface of a cone having its vortex at the element, and a vortex angle of $180° - \phi$, where ϕ is the angle between the cylinder axis and a plane normal to the beam.[6] Experimental verifications of the calculations were made by Farone and Kerker,[13] Cooke and Kerker,[14,15] and Lundberg.[16] The latter used a helium-neon laser producing the 6328 Å line of neon to investigate the scattering by large glass cylinders (6.27 to 49.9 μm in diameter). Figure 4-7 is a diagram of his light-scattering photometer.

The scattering by particles for which $\alpha(m - 1) << 1$ may be calculated with the Rayleigh–Gans approximation. If the inequality is satisfied, the particle need not be small relative to the wavelength. Rayleigh–Gans scattering has been discussed by van de Hulst,[5] who has provided equations that can be applied to scattering by ellipsoids, cylinders of finite length, and randomly oriented rods and disks. For the latter particles, the lobes and sharp minima in the scattering pattern are damped out by the integration over many orientations.

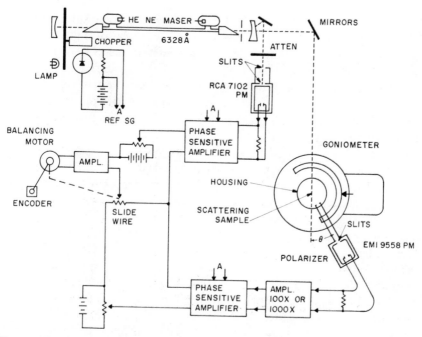

Figure 4-7. Block diagram of light scattering photometer.[16]

The Mie equation applies to single particles and to suspensions of particles that are monodispersed and sufficiently dilute that there is little secondary scattering. However, airborne particles usually have a wide size range and often are of irregular shape. The light scattered by a suspension of such particles is the sum of the scattering by each particle, but the scattering coefficients for the individual particles may vary greatly.

According to Hodkinson,[6,17] a suspension of polydispersed randomly oriented particles will have a resultant scattering pattern similar to that of a suspension of spheres of equal mean volume whose polydispersity in α resembles the orientation polydispersity of the particles. The more the particle shapes are flattened or elongated, the smoother the patterns become. Similarly, the scattering patterns for spheres illuminated with white light will be smoother (the oscillations more damped) than if the illumination were monochromatic. The effects relate both to total scattering (or extinction) as a function of α, and to the variation of the scattering intensity with angle relative to the incident beam. Experiments demonstrating such effects were performed by Hodkinson[17] with powdered quartz, diamond, anthracite, and bituminous coal suspended in water.

Individual airborne particles seldom consist of a single substance, and such particles of a size close to the wavelength of the illuminating light do not scatter the radiation as do homogeneous particles. Kerker and his co-workers[1] have studied scattering by heterogeneous particles, especially stratified spheres, starting with the work by Aden and Kerker.[18] They differentiate between two principal cases: (1) a sphere consisting of two or more concentric layers, each having a constant refractive index, and (2) a sphere for which the refractive index within one or more layers varies continuously but with radial symmetry.

The Mie theory also applies to particles much larger than the wavelength of light, and as the particle size increases the scattering coefficient, Q_{scat}, approaches two (Figure 4-3). Since this coefficient is defined as the effective area for scattering divided by the actual area, a Q_{scat} of two implies that a large sphere scatters twice as much light as falls on it. The explanation is quite simple. The limiting value of two can be satisfactorily explained on the basis of Babinet's principle, which shows that a circular opaque disk diffracts an amount of light around the edges of the disk equal to the amount of light that actually strikes the disk. The amount of light intercepted by a large spherical particle for unit illumination intensity is πr^2, and the particle will also diffract an amount of light πr^2, so the total amount of light intercepted and diffracted is

$2\pi r^2$, corresponding to a scattering coefficient of two. The diffracted light is scattered in a forward direction almost parallel to the incident beam. Confusion has existed concerning the significance of a Q_{scat} value of two, since the light-sensing device (such as a phototube) in a turbidimeter may receive the diffracted light and the apparent value of Q_{scat} will be unity.

1.2.   The Owl

The owl is an instrument developed during World War II for determining the sizes of particles in monodispersed aerosols.[7,19] It is manufactured by the Process and Instruments Corp., Brooklyn, N.Y. The owl is really two instruments in one. Diameters between 0.1 and 0.4 μm are determined from the polarization of the scattered light, while those between about 0.4 and 2 μm are determined from "higher order Tyndall spectra," described below.

When the particles are sufficiently small that Rayleigh's law applies, the polarized component i_2 of the scattered radiation is zero at $\theta = 90°$ (see Figure 4-2). As the particle diameters are increased from the Rayleigh region up to about 0.4 μm, the ratio i_2/i_1 also increases. The owl contains a polarization photometer used to determine this ratio and thus the sizes of particles between 0.1 and 0.4 μm diameter.

As the diameters increase beyond 0.4 μm, the ratio i_2/i_1 passes first through a maximum and then through a minimum, and the polarization photometer is inapplicable. Therefore, another property of the scattered light is used to determine the sizes of the larger particles. When an aerosol whose particles are all of one size and between about 0.2 and 2 μm diameter is illuminated with a beam of white light, a succession of spectra is observed as θ is varied. These result from the variation of scattering with wavelength at various angles relative to the incident beam, and are known as higher order Tyndall spectra. As the observation angle is varied from the forward toward the backward direction, the order of colors is violet, blue, green, yellow, orange, and red. Near 90° the order reverses, and the opposite series is repeated until the backward direction is reached. According to Sinclair and La Mer,[19] the observation is best made through a plane polarizer oriented with its direction of vibration perpendicular to the plane of observation so that only i_1 is seen. The other polarized component, i_2, exhibits a different and usually less distinct series of spectra.

The angular distribution of these effects was calculated with Mie theory by Sinclair[7] and compared with the distributions observed for aerosols prepared from oleic acid, stearic acid, and sulfur. Good agreement was obtained. The purity and brightness of the colors increase with increasing uniformity of the particles. The number of times the series is repeated as θ is varied from near 0 to near 180° increases with increasing particle size. Red is the most distinctive color in the spectra, and when the refractive index of the particles is in the range 1.33 to 2, the number of times red appears divided by 5 approximately equals the diameter in microns. The owl is designed for counting the number of reds as well as for determining i_2/i_1, and the combined techniques cover a particle size range of about 0.1–2 μm diameter.

The owl is a small, portable, instrument the main components of which are a chamber to contain the aerosol, a light source, and a low-power microscope. The aerosol is observed with the microscope through a window in the chamber. The microscope can be rotated to count the number of reds produced by the aerosol. A bipartite polaroid disk is mounted in the ocular of the microscope, which serves as the polarization meter. This disk consists of a plane polarizer with one-half of its plane of polarization perpendicular to a line along a diameter of the disk and the other half parallel to the dividing line, which is placed either perpendicular or parallel to the plane of observation. Another polarizing disk, the analyzer, is placed between the bipartite disk and the observer. A pointer is attached to the analyzer and a scale reading in degrees is mounted on the microscope tube behind the pointer. When the plane of vibration of light passing through the analyzer is in the plane of the observer, the pointer indicates zero. The size of particles in the smaller of the two size ranges is determined by observing the aerosol in the chamber through the microscope and turning the analyzer until the intensities of light through the two sides of the bipartite disk are matched. The ratio i_2/i_1 is then calculated from the angle ϕ indicated on the scale:

$$\frac{i_2}{i_1} = \tan^2 \phi \tag{4-6}$$

Calibration curves for the polarizer function of the owl, which eliminate the need to use equation (4-6), are presented in Figure 4-8. They were calculated for monodispersed, transparent droplets.

Advantages of the method are the rapidity with which measurements can be made and the small particle sizes for which it is applicable.

Kerker and La Mer,[20] working with sulfur hydrosols, extended the polarization ratio approach to values of α between 3.5 and 10. Presumably the method could be applied to aerosols. For particles in this range,

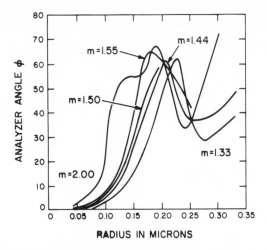

Figure 4-8.   Calibration curves for the polarization photometer function of the owl when $\theta = 90°$ for varying refractive indexes of the particles.

i_2/i_1 oscillates with varying θ, and the method involves measuring the angular location of the maxima and minima.

Maron and Elder[21] studied the higher-order Tyndall spectra exhibited by butadiene-styrene latex spheres ranging in diameter from 0.26 to 1.01 μm. They showed that the relationship between the diameter d and θ of the first red or green orders is given by

$$d \sin (\theta/2) = k' \tag{4-7}$$

where k' is a function of the color and of the refractive index ratio of the particles to the suspending medium. According to the authors, the equation is valid for both mono- and polydispersed particles, and yields weight-average diameters.

Maron, Pierce, and Elder[22] made a similar study of the use of the polarization ratio (i_2/i_1) for determining particle sizes, based on the angular dependence of the first two minima and maxima of i_2/i_1 on particle size. They derived the equations

$$\frac{d}{\lambda} \sin\left(\frac{\phi_i}{2}\right) = l_i \tag{4-8}$$

and

$$\frac{d}{\lambda} \sin\left(\frac{\omega_i}{2}\right) = L_i \tag{4-9}$$

where ϕ_i and ω_i are the angles at which the minima and maxima appear, respectively, and i ($=1,2,3$, etc.) is the order in which they occur. The constants l_i and L_i for any one order depend only on the ratio of the refractive index of the particles to that of the suspending medium. Maron et al. concluded that, depending on the wavelength employed, the first two minima and maxima can be used to determine particle diameters from about 0.13–6.3 μm.

1.3.   Integrating Nephelometers

A nephelometer is an analytical device for determining the concentration of particles suspended in a liquid or gas by measuring the intensity of scattered radiation. An integrating nephelometer, as the name implies, integrates the flux of the scattered radiation over a wide range of θ. In principle, such a device measures the extinction coefficient, b_{scat}, of the Beer–Lambert Law and includes the scattering by both molecules (Rayleigh scattering) and particles:

$$b_{scat} = b'_{scat} \text{ (particles)} + b''_{scat} \text{ (molecules)} \qquad (4\text{-}10)$$

Also,

$$b_{scat} = 2\pi \int_0^\infty \beta'(\theta) \sin \theta \, d\theta \qquad (4\text{-}11)$$

where $\beta'(\theta)$ contains all of the angular dependence of the scattering.

According to Butcher and Charlson,[23] the first such instrument was devised by Beutell about 1943 (see Beuttell and Brewer[24]). Since then numerous modifications have been made, and the one most used today was devised by Ahlquist and Charlson.[25,26,27] Their device measures extinction coefficients down to less than 1×10^{-6} m^{-1}, a sensitivity sufficient to permit the use of filtered gases such as air, carbon dioxide, or Freon 12 as light-scattering standards.

The main elements of the Ahlquist-Charlson nephelometer are shown in Figure 4-9 and a commercial model is shown in Figure 4-10. No lenses or mirrors are used in the construction. The light source which illuminates the sample is a xenon flash lamp and a 5-cm diameter opal glass diffuser to provide a cosine emission characteristic. The detector is a ten-stage photomultiplier tube that receives scattered light that passes the collimator and light trap. The cone of observation is determined by the first and fourth disks. The fifth disk in the collimator and those in the light trap cast shadows on any surfaces reflecting light to the photo-

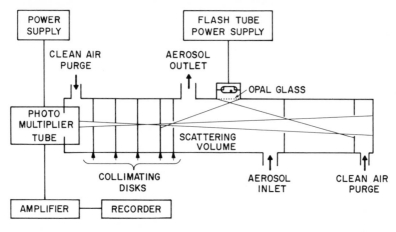

Figure 4-9. Integrating nephelometer. Reprinted with permission from *J. Air Poll. Control Assoc.* **17**, 467 (1967).

multiplier. All inside surfaces are painted black. A four-wavelength version has been designed to provide information concerning the size distribution of the particles in the sampled air. As can be seen from Figure 4-9, such nephelometers do not collect light scattered from the entire range of θ (0–180°). The usual range is from about 8°–170°. However, the error for the earth's atmosphere, polluted or not, is usually 10% or less.[23]

Integrating nephelometers, including the Ahlquist-Charlson version, were originally developed to provide an indirect measurement of visual range, which can be defined as the distance at which the contrast between a black object and the adjacent sky becomes equal to ϵ, which is the least fractional difference in intensity the eye can detect. The relationship between b_{scat} and visible range, L, is provided by the empirical equation

$$L = \frac{3.9}{b_{\text{scat}}} \tag{4-12}$$

which is based on the assumption that the contrast limen, ϵ, sometimes called the psychophysical constant, is 0.02.

However, during recent years the integrating nephelometer has been used mainly to measure mass or volume concentrations of particulate material in the atmosphere, especially in smog. Such an application, of course, requires a calibration with the airborne particles, and this appli-

Figure 4-10. Integrating nephelometer. Courtesy of Meteorology Research, Inc., Altadena, Calif.

cation to a particular atmosphere must be based on the assumption that the size distribution and composition of the particles in the sampled atmosphere is similar to those of the particles used for calibration.

Charlson et al.[28] have discussed this application of the instrument. Figure 4-11 shows the dependence of b_{scat} (m^{-1}) on the volume of aerosol particles calculated from measured size distributions; it indicates that

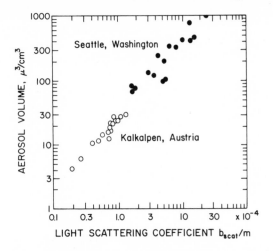

Figure 4-11. The dependence of scattering coefficient (m^{-1}) on volume of aerosol particles $(\mu m^3/cm^3)$ calculated from measured size distributions. Note: The solid circles are based on Seattle, Washington, data and the open circles on Kalkalpen, Austria, data.[28] Source: National Air Pollution Control Administration (NAPCA) (1969).

b_{scat} is directly proportional to the volume of suspended particles. They compared mass concentrations of particulate material collected with glass fiber filters with b_{scat} in Seattle, Washington under widely differing haze conditions. The results are shown in Figure 4-12.

Charlson, Ahlquist, and Horvath[29] extended this work by making a series of simultaneous mass concentration and light-scattering (nephelometer) measurements in several locations with widely varying types and amounts of suspended particulate material. They concluded that the relationship between mass concentration and b_{scat} is the same for all locations. However, the scatter was quite large. They also found that the most probable value of the product of visible range and mass concentration is about 1.2 g m^{-2}, with 90% of all cases being between 0.7 and 2.6 g m^{-2}.

Thielke et al.[30] measured light scattering in Los Angeles smog with two broad-band and one multiwavelength integrating nephelometers. They found that wide-band light scattering correlated very highly with monochromatic light scattering. A least-squares fit to a power-law size distribution was made, and the exponent was found to correlated highly with the exponent of a power-law approximation of the wavelength dependence of scattering.

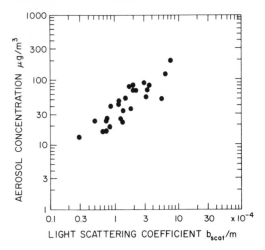

Figure 4-12. Measured dependence of mass of aerosol particles per unit volume ($\mu g/m^3$) on the light-scattering coefficients (m^{-1}) in Seattle, November-December 1966.[28] Source: NAPCA (1969).

1.4.   Particle Counters and Sizers

During and immediately following World War II, the first devices were built to count and determine the sizes of individual airborne particles in situ by light scattering without image formation. Several of the earlier devices produced photographic records of the diffraction patterns produced by the particles. Examples were the Chaney dust camera[31] and a similar device developed by LaMer and Lee.[32] Other instruments based on this principle used photoelectric counters instead of photographic techniques to display or record the output. Gucker, O'Konski, and their co-workers were pioneers in this work.[33,34,35] The early Gucker and O'Konski devices counted particles but did not size them. A typical arrangement, shown in Figure 4-13, demonstrates the principles involved. A narrow stream of aerosol, protected by a flowing sheath of pure air, passed through a cell illuminated by a beam of light. The aerosol had to be so dilute (or the stream so narrow) that only one particle at a time was illuminated. Each particle produced a pulse of scattered light and a photocell received the forward-scattered light. The pulses were counted electronically to yield the particle number concentrations in the aerosol. Instruments were also developed using light scattered at other angles.

For a considerable size range, the total amount of light scattered by each particle increases monotonically with increasing particle size. Thus,

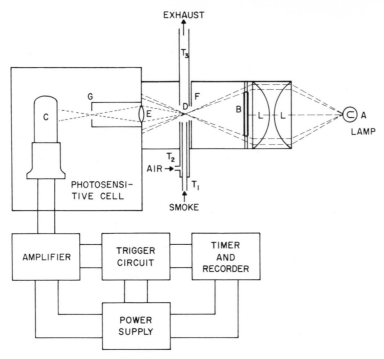

Figure 4-13. Schematic diagram of photoelectronic particle counter. Reprinted with permission from Gucker and O'Konski, *Chem. Revs.* **44,** 373 (1949). Copyright by the American Chemical Society.

by incorporating a pulse-height discriminator in an instrument such as that shown in Figure 4-13, size distributions as well as number concentrations can be obtained. Such instruments must be calibrated with particles of known size and optical properties.

One of the first instruments based on this principle was developed at the Armour Research Foundation (now the Illinois Institute of Technology Research Institute, IITRI) and was designed for relatively large particles, 1–64 μm diameter.[36] This instrument, called the aerosoloscope, made use of light scattered at θ = about 90° (right-angle scattering). Particles were counted at a maximum rate of 2000 per minute, and the pulse amplitudes sorted in 12 channels.

Royco Instruments, Inc., manufactures a wide range of instruments that use right-angle or near-forward scattering.[37] For example, their Model 200 is programmable for sequential automatic scanning of 15 discrete particle size ranges (0.3–>10 μm diameter) with three selectable

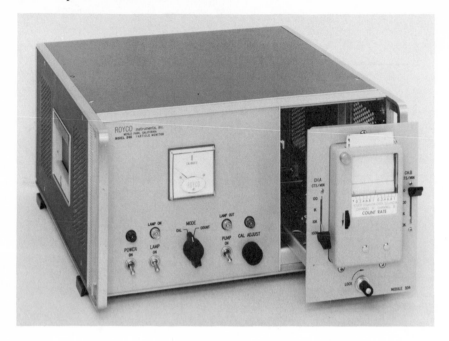

Figure 4-14. Royco particle counting and sizing instrument, Model 245 equipped with Module 504. Courtesy of Royco Instruments, Inc.

scan periods. In order to prevent the counting of more than one particle as a single particle, the air sampled must often be diluted. The Model 202 dilution system is incrementally adjustable from 0 to 10:1. The right-angle scattering systems are recommended for maximum size resolution, while the forward-scattering systems are recommended for monitoring large quantities of ambient air in which the suspended particles vary widely in composition, size, and transparency.

Figure 4-14 shows the Royco Model 245 equipped with module 504. Model 245 uses near-forward scatter at a flow rate of 1 ft^3 min^{-1}. Seven modules are available. Module 504 is a strip-chart recorder having two pens and thus reads two size ranges simultaneously. Another module for this model is shown in Figure 4-15. This one (507) produces a digital display, and has a five-channel memory for five size ranges.

Bausch and Lomb manufactures a single-particle counter using essentially forward light scattering and a 40° light-acceptance angle by the photomultiplier. It is designed for the diameter range 0.3–10 μm. The B&L Dust Counter 40-1 is equipped with a size-selector switch that can

Figure 4-15.   Digital display module for use with the Royco Model 245 particle counting and sizing instrument (Figure 4-14). Courtesy of Royco Instruments, Inc.

be set at one of several size ranges. Brown[38] has described circuitry that can be added to this counter to allow unattended, automatic measurements from which size distributions can be determined. The modified system has been flown on aircraft to altitudes of 20,000 ft.

The Climet Instrument Co. manufactures several airborne particle counters that use near-forward scattering with a 60° acceptance angle. Their model 201, Figure 4-16, like the B&L instrument, is designed for the diameter range 0.3–10 μm.

TechEcology, Inc., manufactures a near-forward scattering optical system with a solid-state detector, unitized modular construction, and a solid-state calibration system. Depending on the method of recording or display, it can indicate up to five size ranges from >0.5 μm to >10 μm diameter. They also sell an aerosol generator for the calibration of light-scattering particle analyzers that atomizes aqueous suspensions of monodispersed polystyrene particles.

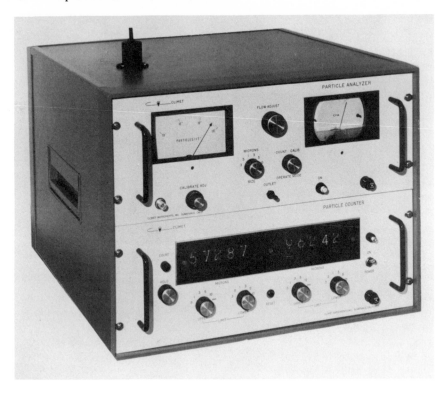

Figure 4-16.   Climet Instruments Co. Model 201 particle counter. Courtesy of Climet Instruments Co.

The Sinclair (Phoenix Instrument Corp.) instrument has one of the widest size ranges (0.3–>16 μm diameter and an especially large sampling rate (2800 cm³ min⁻¹).

Other manufacturers of such light-scattering systems are the Dynac Corp., which produces a portable version, and Coulter Electronics, which markets a counter for monitoring particles in controlled environments.

Rosen[39] has designed a counter to be flown into the stratosphere on balloons. It is a two-channel device, one channel counting all particles greater than 0.55 μm diameter and the other counting all particles greater than 0.70 μm diameter. It has been flown successfully for many years in many parts of the world for determining the concentrations of airborne particles.[40,41,42]

Hodkinson and Greenfield[43] made theoretical calculations of the behavior of such counters. They pointed out (as mentioned earlier) that the use of white light alone smooths the curves of collected flux versus particle size, and little further smoothing is produced by the use of large-aperture illuminating and collecting lenses. However, the smoothing does not eliminate the effects of the optical properties of the particles. They suggest that $\theta = 40°$ to $50°$ is best for transparent particles and that, in instruments measuring the forward lobe of the scattering pattern, the differences in response for transparent and absorbing particles are least but not negligible.

Quenzel[44,45] has undertaken a theoretical study of the influence of the refractive index of particles on the accuracy of size determination with light-scattering aerosol counters. He concluded that high resolution of the aerosol size distribution is impossible, particularly because of the different refractive indices of the atmospheric aerosol particles. For certain counters and refractive indices the response of the instruments does not increase monotonically with increasing particle size, according to Quenzel's calculations.

A number of experimental studies have been made of the response of single-particle counters to particles of various types, many in the course of calibrating the instruments. One of the most complete was by Whitby and his co-workers,[46,47] who include a bibliography and table of calibration methods used by various investigators. Three counters were used: the Royco PC 200, a Bausch and Lomb prototype counter, and a counter designed by the Southern Research Institute and described by Thomas et al.[48] The aerosols were chosen to contain nonideal but nearly mono-dispersed particles. The aerosols were produced with the spinning disk generator described by Whitby et al.[49] These aerosols were polystyrene latex spheres, dioctyl phthalate, India ink, and polystyrene. The sizes of the particles were determined by light and electron microscopy.

The resolving power in terms of the standard deviation (s_g) for all three counters for the polystyrene latex particles was essentially the same, decreasing from about 1.4 at 0.36 μm to between 1.15 and 1.2 at 1.3 μm. The authors suggested that the decreasing resolving power for the submicron-diameter particles may be due to the combined effects of instrument noise and a background count of submicron particles from the aerosol generator. The resolving power for all three counters for dioctyl phthalate aerosols in the range from 1–5 μm was quite good $(s_g = 1.1$ to $1.26)$. The s_g for the microscope determinations was about 1.1. Highly variable values of s_g were obtained for the India ink and polystyrene aerosols, probably because the particles were quite irregular. Each of the polystyrene particles had a dent, and a few of the smaller particles were hollow.

Another problem that may arise when using single-particle counters is that spurious counts may be produced if large numbers of particles that are too small to be counted are present in the counter sensing zone. This problem can arise, for example, when polystyrene aerosols are generated by "atomizing" suspensions of the polystyrene spheres unless the suspending medium is free of any solute, since the evaporation of droplets containing no spheres will produce extremely small particles as residues. Whitby et al.[47] developed a statistical theory that can predict the size distribution and number of the spurious pulses from the concentration and particle size of the "subcountable" aerosol. They extended the theory to atmospheric aerosols obeying the distribution law

$$\frac{dN}{d(d)} = 0.41 \ vd^{-4} \qquad (4\text{-}13)$$

where v is the volume of particles per unit volume of air.

A problem which Whitby[47] mentions in connection with the Royco PC-200 is that the optical system, in addition to the sampling region, fills with aerosol as the aerosol is sampled. When clean air is introduced into the sampling system after the instrument has been measuring the particles in an aerosol, the particles in the optical system continue to give a count for a long time. The problem is best eliminated by providing a clean sheath of air around the aerosol. Whitby designed a new inlet system with such a sheath for the Royco PC-200.

Rimberg and Keafer[50] compared particle sizes and concentrations determined with the Royco 220 and the Bausch and Lomb 40-1 counters with the number concentrations determined by collecting the particles on membrane filters and counting them with a microscope. The Royco 220 was set for the size ranges 0.5–5 μm and $\geq$ 5 μm, whereas the B&L was set for $\geq$ 0.5 μm. The test aerosol particles were monodispersed polystyrene or polyvinyltoluene latex spheres 0.500, 0.81, 1.099, and 1.947 μm in diameter. The results showed that both counters set at $\geq$ 0.5 μm were inaccurate for particles smaller than 1.0 μm, with substantial proportions of the particles not being measured. Rimberg and Keafer suggested two possible explanations: that the illumination of the optical sensing region may not be uniform and that the photocathode surface may not be uniformly sensitive.

In spite of the limitations described above, optical counters have been found to be extremely valuable for investigating aerosols of various types, including polluted and natural atmospheres.

1.5. In-situ Photomicrography

The microscopic examination of particles usually involves a preliminary collection. However, photomicrographs of particles can be taken while the particles are still airborne. For the most part, equipment built for in-situ photomicrography is designed for quite large particles, such as cloud and spray droplets. However, particles as small as 0.5 μm radius have been photographed in this manner.

The requirements for the direct photomicrography of airborne particles include those for the photomicrography of particles collected on microscope slides, that is, a microscope, a camera body and film holder, and a light source. However, two characteristics of airborne particles add complications not encountered during conventional photomicrography. These characteristics are (1) the particles are in motion, and (2) the number of particles in focus at any given time is generally very small when a microscope or similar objective is used. Therefore, there may be only a small chance that an aerosol particle will be in focus at any given time. The limitations with regard to the volume in focus are readily derived from Table 3-1. If we assume a film size of 5 × 7 in. and a magnification of 250X the numerical aperture, the volume in focus for numerical apertures of 0.1, 0.25, and 0.85 will be 0.99, 0.024, and 0.0001 mm^3, respectively. If an aerosol number concentration is 10^4 particles cm^{-3}, there would be only one chance in 1000 that a particle would be in focus at any given moment for a numerical aperture of 0.85.

Two general methods have been used to photographically "stop" the particles. If the aerosol is moving rapidly past the photographic equipment, the motion of the image across the photographic film can be slowed or stopped with a rotating mirror or prism. If the particle motion is more or less random, very short exposure times with high-intensity flashes may be required, especially at high magnifications.

Figures 4-17 and 4-18 show schematically the optical components of a photomicrographic system used to photograph ice fog crystals in Alaska.[51] Very short exposure periods (1–10 μsec) and high-intensity illumination (about 2.5 × 10^4 lumen-seconds per flash) were employed. Dark-field illumination was provided with a flash tube and suitable lens system. A large number of flashes was used to expose each frame of film so that a number of images of particles was formed on each negative. The large number of flashes permitted the use of objective lenses having relatively high resolving powers but correspondingly small depths of field. Furthermore, the flash repetition rate was high (up to 100 flashes

258

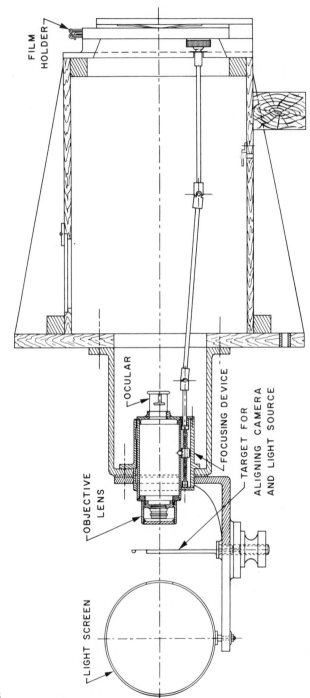

FILM HOLDER

C-41520-19

OCULAR

FOCUSING DEVICE

TARGET FOR
ALIGNING CAMERA
AND LIGHT SOURCE

OBJECTIVE
LENS

LIGHT SCREEN

Figure 4-17. Microscope and camera body of system for in-situ aerosol particle photomicrography. From Cadle and Wiggins
A.M.A. Archives of Industrial Health **12**, 584 (1955). Copyright 1955 by American Medical Association.

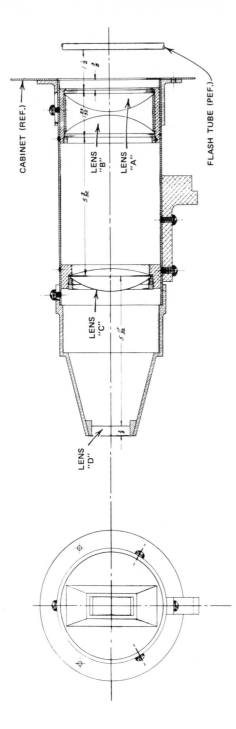

Figure 4-18. Optical system for condensing light from the flash tube for in-situ aerosol particle photomicrography. Dimensions are in inches. From Cadle and Wiggins, *A.M.A. Archives of Industrial Health* **12**, 584 (1955). Copyright 1955 by American Medical Association.

per second), so that numerous exposures could be obtained on each negative within a reasonable time.

The body-tube length of the microscope shown in Figure 4-17 was the conventional 160 mm, and the drawing is scaled according. Various objectives were used, including microscope objectives and high numerical aperture camera lenses. The latter were used in a reversed position to take advantage of their corrections for aberrations because of the short object-to-lens distance used in photomicrography. The most useful lenses were a 0.25-N.A. microscope objective and a movie-camera lens. The ocular was a 6X Huygenian. These combinations provided sufficient magnification that the limit of resolution of the objective, rather than the limit of resolution of the film, controlled the resolution obtained.

The camera was provided with a ground glass screen that was used to focus the microscope on a "target" placed in the beam of light from the flash tube. The target was a piece of rod, flat on one side, which was removed after the microscope was focused.

Airborne particles passed through the field of view of the microscope by natural convection of the air. Extraneous light was prevented from entering the objective lens by a light trap consisting of a hollow ball with a circular hole facing the objective that was painted black on the inside and around the rim of the hole. The hole was large enough to include the cone defined by the lens aperture and focal point. No shutter was used, the only shuttering action being provided by the flash itself. The flash tube was a gas-discharge tube about 20 cm long and 0.6 cm o.d., shown in Figure 4-18. An optical system, also shown in Figure 4-18, produced a narrow light beam that illuminated the field of view of the objective. The lenses were all spherical except the one farthest from the flash tube, which was a cylindrical lens ground from a short piece of borosilicate glass rod.

Suspensions of transparent droplets were found to produce somewhat misleading images. When such a droplet was illuminated from the side (the usual mode of illuminaton with this equipment), two small parts of the droplet, at opposite ends of the diameter parallel to the illuminating beam, were very much brighter than the rest. Thus the image of each droplet consisted essentially of two small dots, and the faint image of the rest of the droplet was easily overlooked.

Cannon[52,53] has developed photomicrographic equipment for water droplets and ice crystals. A modification for mounting on a sailplane, which can produce images of droplets and particles larger than about 4 μm diameter, is shown schematically in Figure 4-19. The airfoil housing mounted above the glider is shown in Figure 4-20.

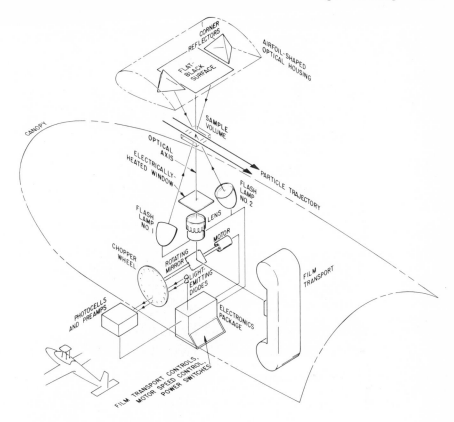

Figure 4-19. Schematic of Cannon particle camera. From T. Cannon, *Rev. Sci. Inst.* **45**, 1448 (1974).

The photomicrographic equipment is in two parts. The airfoil-shaped optical housing shown in Figure 4-19 contains two corner reflectors and a flat-black background for the photomicrographs. Inside the cockpit are the film transport, lens, rotating mirror, controls, and electronics. The flash lamps are located on either side of the single lens (this is a simple rather than compound microscope), and produce a 10-μsec pulse of light measured at one-third peak intensity. The region in focus is illuminated by light coming directly from the lamps and by light from the corner reflectors. Each corner reflector is a tetrahedron with mirrors on three sides and a glass window on the side facing the lamps.

The lens has a focal length of 135 mm and produces one-to-one images, that is, no magnification or minification. Thus considerable enlargement of the photomicrographs is required.

Figure 4-20. Airfoil-shaped optical housing for the Cannon particle camera mounted on a glider.

In-situ photomicrography has been used by the author to photograph highly volatile droplets formed by an explosion. Since the droplets were about 20–30 μm diameter, a low numerical aperture lens having a relatively large volume in focus could be used as the objective. The illumination was transmitted instead of dark-field. A single exposure was made, firing a flash tube milliseconds after the explosion, with a time-delay mechanism built into the electronic circuit that fired the explosive. Because only one flash was used, power requirements were modest and were satisfied with three 300-V batteries and a capacitor. High-speed Polaroid film was used; this film is, of course, particularly useful for field operations. The entire system was rigidly mounted on a single steel plate to avoid misalignment produced by the shock wave.

1.6. Holography

Closely related to in-situ photomicrography is holography.[54] One of the first, if not the first, practical applications of holography was a device to investigate rapidly moving airborne particles in the size range 3–3000 μm diameter. It was designed to provide images of all the particles in a single volume simultaneously and with equal sharpness. Thus it was especially useful for recording the images of snowflakes, raindrops, and cloud droplets. A modulated ruby laser of 10 mW power illuminated the particles in a volume as large as 5000 cm³ for 20 μsec. The hologram recorded the three-dimensional distribution of particles that later could be viewed with a continuous-wave laser.

Koechner has described equipment for the holography of aerosol particles emerging from a nozzle.[55] The equipment is shown schematically in Figures 4-21 and 4-22. As usual, the laser beam pulse is divided into

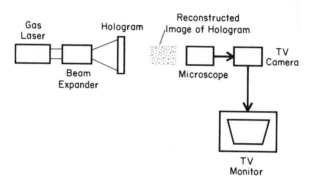

Figure 4-21. Equipment for preparing holograms of aerosol droplets emerging from a nozzle.[55]

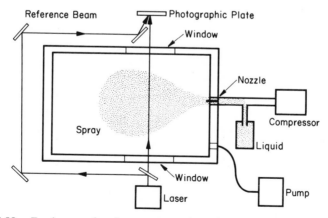

Figure 4-22. Equipment for the reconstruction of a hologram of airborne droplets and studying them with a microscope and TV monitor.[55]

an object beam and reference beam. The former is directed through a window into a chamber containing the aerosol and out of a window on the opposite side. The light scattered in a forward direction by the droplets produces an interference pattern with the reference beam which is recorded on the photographic plate to form the hologram. The image reconstructed in space in three dimensions can be magnified with a microscope and displayed by closed circuit television. By moving the hologram, the entire volume sampled can be studied.

1.7.   Remote Sensing

The methods described thus far for particle measurement have involved collecting the particles or measuring them in-situ but in a relatively small volume in or adjacent to the instrument. However, a number of methods based on light scattering can be used to investigate the nature of suspended particles at large distances from the equipment. The need for such techniques is obvious, and the suspensions range from the smoke emitted from stacks to galactic dust. The technique may be either active or passive. An active technique is one in which the sensing radiation, such as a beam of light, is produced by the measuring system, whereas a passive technique uses radiation already existing. A wide variety of platforms may be used, including the ground, aircraft, balloons, satellites, and rockets. An excellent review of the remote sensing of the troposphere has been edited by Derr.[56]

1.7.1.   Laser Radar

Laser radar or lidar (which is an acronym for light detection and ranging) is being used extensively for the remote detection of airborne particulate material, determining the distance from the instrument of the particulate matter sensed, and obtaining semiquantitative information concerning the concentrations of particles. In operation, a pulse of coherent light is emitted from a laser which at any moment illuminates a well-defined volume of air. The molecules and airborne particles backscatter part of the light, which is collected by a receiving telescope. The collected light is passed through a narrow-band interference filter to decrease the intensity of background light from the sky and directed onto the face of a photomultiplier tube. The amplified signal from the photomultiplier can be displayed as a function of time on an oscilloscope. The time interval between the production of a laser pulse and the return of light scattered from the pulse is used to calculate the distance of the scattering particles. The intensity of the scattered light provides a rough indication of the quantity of particulate material in the illuminated volume. This subject has been reviewed by Collis.[57] Various lasers can be used, such as Q-switched ruby or neodymium-doped glass systems, producing pulses with peak powers of tens of megawatts and durations of 10–20 nsecs. Fox et al.[58] have described an airborne laser radar designed to measure stratospheric aerosols. A dye laser is used, consisting of a flash lamp that provides the energy source for lasing action and Rhodamine-6G dye dissolved in methyl alcohol.

In order to estimate the amount of light backscattered by the particles in a given illuminated volume, one must estimate the amount of light backscattered by the molecules in that volume. The results are usually reported as the scattering ratio, defined as the ratio of observed to molecular (Rayleigh) scattering cross sections. The Rayleigh scattering is usually estimated from values of molecular number density given in standard atmospheres. A better method is to use two wavelengths, the shorter of which is strongly scattered by the air molecules. Sometimes the molecular scattering is estimated from radiosonde measurements of temperature and pressure. The attenuating properties of the aerosol particles can usually be ignored, but attenuation due to the molecules and to ozone should be included in the solution.

1.7.2. Coronae

When a distant light source is observed through a suspension of particles that are nearly all the same size, diffraction rings (coronae) are seen surrounding the source. Such rings are often observed around the sun or moon when the atmosphere contains suspended water droplets, ice crystals, or dust of appropriate size distribution and concentration. The mean size of the particles can be estimated from the angle θ', formed by the light source, the observer, and one of the rings. The particles must be large compared with the mean wavelength of the light observed, but not so large that the diffraction rings are indistinguishably crowded together.

The distribution of rings produced by a single opaque sphere that is large compared with the wavelength of light is described by Fraunhofer diffraction theory. The distribution of rings for a suspension of particles all of one size is the same as that for a single particle. The equation describing the position of the first bright ring is $\sin \theta' = 0.819 \, \lambda/r$; for the second it is $\sin \theta' = 1.346 \, \lambda/r$; for the third it is $\sin \theta' = 1.858 \, \lambda/r$;

and for the fourth it is sin $\theta' = 2.362\ \lambda/r$. The angular positions of the dark rings are given by the equation:

$$\sin \theta' = \frac{n + 0.22\lambda}{d} \tag{4-14}$$

where n is the order of the rings.

Although these relationships were developed for opaque particles, they apply to transparent particles for which α $(\pi d/\lambda)$ is greater than about 30. The particles need not be as nearly monodispersed as is required to observe higher-order Tyndall spectra. Sinclair[7] observed coronae produced by water fogs that contained droplets varying in radius from 4–16 μm or greater. He also compared the sizes of lycopodium spores as determined with a microscope and as determined from the angular positions of light and dark diffraction rings produced by a suspension of the spores in air. The former method yielded a range of radii of 13.6–16.0 μm, whereas, the latter yielded a mean radius of 15.5 $\pm$ 0.2 μm. Van de Hulst[5] concludes that there is no conspicuous difference between the coronae produced by random needles and those produced by spheres. On the other hand, van de Hulst has stated that anomalous diffraction can cause difficulties in interpreting coronae, and that the intensity of the first bright ring and the position of the first dark ring are subject to strong fluctuations for $\alpha < 50$.

Coronae have been used extensively for particle size determinations, for example, of fog droplets. An interesting application is the conclusion by van de Hulst[5] that the dust in space producing the zodiacal light must have a wide size distribution since diffraction rings are not observed.

1.7.3.   Polarization Measurements

The polarization of light scattered by remote suspensions of particles, such as the clouds of planets other than earth and galactic dust, has often been used to provide information concerning the composition and the size distribution of the particles. For example, the light from a distant star has been found to be polarized only if it is reddened, presumably by galactic dust, and the polarization is independent of the physical properties of the star. Thus the polarization is also most likely produced by the galactic dust. The degree of polarization can be explained if the particles are elongated and oriented. Possibly the particles are ferromagnetic and are oriented by magnetic lines of force, along the spirals of the galaxy.[59-64]

Lyot[65] compared a plot of the degree of polarization of the light reflected by Venus versus scattering angle with similar plots for water droplets. He suggested that the mean diameter of the droplets in the Venus clouds is about 2.5 μm. A much more recent study of the nature of the Venus clouds as derived from their polarization was undertaken by Hansen and Hovenier.[66] They analyzed the linear polarizaton of sunlight by comparing observations with extensive multiple-scattering computations. They concluded that the cloud particles have a mean diameter of about 2 μm, a narrow size distribution, and probably consist of a concentrated solution of sulfuric acid.

1.7.4. Ringelmann Charts

A method used for many years by "smoke inspectors" for estimating the density of smoke emitted by stacks involves the use of Ringelmann charts. They consist of arrays of black lines, crossing at right angles, on white backgrounds such that the charts appear grey when placed at some distance from the observer. The proportion of black to white areas is varied so that the charts produce a graduated scale of grey in five steps. In use, the chart is hung at eye level about 50 ft from the observer, who compares the smoke with the charts, and selects the chart which most nearly corresponds to the shade of the smoke. The chart number indicates the smoke darkness. An experienced observer can eliminate the charts, merely giving a Ringelmann number to the smoke density.

Although there are many fundamental difficulties with such a system, it has been used extensively for air pollution control in the United States. An obvious difficulty is that the emissions from many stacks are light colored or even white. Furthermore, the optical density of plumes containing deliquescent particles or impure water droplets will vary markedly with the relative humidity and density of the ambient air. Some air pollution control districts estimate the densities of light-colored plumes, assigning them numbers corresponding to the Ringelmann numbers.

Small cards that are miniature replicas of the Ringelmann charts are sometimes used, as are photographs of plumes of varying optical density or reflectance. Translucent slides that are viewed against the sky sometimes are used for the same purpose as the Ringelmann charts.

The U.S. Environmental Protection Agency has demonstrated that plume opacity can be estimated accurately at night with a night vision instrument similar to the army device called the Starlight-Scope. The increase in light intensity is of the order of 50,000 times, which allows daylike viewing of scenes otherwise too dark to see.

Numerous instruments have been constructed for estimating the shade of grey of smoke plumes with the aim of obtaining more precise values than is possible by direct comparison with Ringelmann charts. An example is that of Rose, Nader, and Drinker,[67] which consists of a sighting tube and a very sensitive photometer, equipped with a collimating tube to limit the light reaching the photometer to that scattered and transmitted by the smoke plume.

1.7.5.   Color Effects

Colors of the sunset, sun, and moon have at times been used to estimate the mean sizes of particles in the atmosphere and the altitudes at which layers of particles occur. The color effects have been particularly useful when supplemented by measurements of the actual spectra. The colors are especially noticeable when huge amounts of particulate matter have been injected into the atmosphere, as by forest fires, dust storms, and volcanic eruptions.

Dave and Mateer[68] have reviewed the literature dealing with the effect of stratospheric dust on the color of the twilight sky. They evaluated the "chromaticity of the scattered radiation" received by a terrestrial observer from differing parts of the sky during twlight for six different model atmospheres.

1.8.   Miscellaneous Methods

In addition to the methods described above for measuring particles by light scattering, a large assortment of methods have been used or proposed, mainly for special situations. Perhaps the best known of these is the dissymmetry method, which has been used mainly to investigate macromolecules suspended in liquids. However, it is also applicable to aerosols. For particles in the diameter range 0.1–2 μm there is a progressive change from a symmetrical to a forward-directed scattering pattern. The method involves measuring the intensity of the scattered monochromatic light at two angles that are symmetrical about $\theta = 90°$. Kerker[1] has calculated and plotted the dissymmetry ratio intensity at $\theta = 45°$/intensity at $\theta = 135°$ versus L/λ for spheres, coils, and rods, where L is the diameter of the sphere, the root-mean-square distance between the ends of the coil, or the length of the rod. The method is most useful when the size is just beyond the range of validity of the Rayleigh equation. When the particles are much larger, the effect of shape increases, and the curve for spheres becomes very steep. Instruments for making such measurements are manufactured by the Phoenix Instrument Co., Philadelphia.

Leitz manufactures a device designed to measure dust concentrations where they may be especially high, as in many mines. This instrument, called the Tyndalloscope, measures the light scattered about the angle $\theta = 30°$ from a beam from a small lamp. The scattered light illuminates half the field of an eyepiece. The other half is illuminated directly by light from the same lamp, which passes through a polarizer and analyzer. A measurement is made by rotating the analyzer until the illumination on both sides of the eyepiece is equal. The amount of rotation indicates the dust concentration when the instrument has been calibrated with the type of dust measured. Hodkinson[6] states that for application in coal mines, this instrument's greater response to transparent than to black particles makes it especially responsive to the quartz-containing particles that constitute a great pneumoconiosis hazard. Thus its use in coal mines is more appropriate than instruments based on the measurement of the darkening of filter papers by the dust.

Kerker et al.[69] have developed a method for determining the size distribution of spheres that uses the polarization of the scattered monochromatic radiation for various values of θ. The intensities of the i_1 and i_2 components are measured at a number of angles, and the ratios of the intensities for each of the several angles used to determine the size distribution. An advantage to using ratios is that instrument readings can be used directly instead of having to be converted to absolute intensities. To apply this method, an assumption must be made concerning the shape of the size distribution, and Kerker et al.[69] assumed that it could be desscribed by the ZOLD, a type of size distribution function described in Chapter 1. Theoretical computations of the ratio were made based on Mie theory for three refractive indices, 19 values of θ, numerous values of the mode α_{mo} of the size parameter $\alpha(=2\pi r/\lambda)$, and numerous ZOLD standard deviations. The calculations were made with an IBM 7090 computer and comprised 382,536 cases. The results were stored on punched cards. The ZOLD is completely determined by the values of the mode and of the ZOLD standard deviation. These values were determined for a given aerosol by comparing the computed ratios with the experimental ratios for each of the nineteen values of θ, using an IBM 1620. Solutions consisted of pairs of values of the ZOLD parameters for which agreement was obtained between the experimental and theoretical values of the ratios for all or nearly all values of θ to within a tolerance of 12%.

Other methods for obtaining size distributions that have been proposed include the measurement of the light scattered at a particular angle from an aerosol undergoing differential settling,[70] measurement of the angular variation of the intensity of forward-scattered light at very small angles and at a single wavelength,[71] and determining the intensities of each of the two polarized components of the scattered light as a function of the scattering angle.[72] The use of ultramicroscopes for determining mean particle size was discussed in Chapter 3. Ultramicroscopes are seldom used today, but there are exceptions, such as the stereo-ultramicroscope described by Wilson and Cavanagh.[73]

An unusual application of light-scattering that is being used in experimental studies of nucleation and aerosol coagulation is called laser Doppler spectroscopy. Particle sizes are estimated from their Brownian motion, which in turn is determined from the Doppler shifts produced by that motion in the wavelength of the laser radiation. Two general techniques have been used. In the heterodyne method the scattered light is mixed or "beat" with the original unshifted laser light in a photomultiplier tube. This approach measures absolute motion. Homodyne detection involves allowing each element in the optical spectrum to beat with every other element, and measures the motion of particles relative to each other.

2. LIGHT EXTINCTION

2.1. Theory

The amount of light that passes through a given length of aerosol is, of course, the difference between the amount of incident light and the amount scattered and absorbed by the particles. Thus the theory is very similar to that for light scattering, but there are a few differences.

The Lambert-Beer law (which Hodkinson[6] called the Bouguer law since it was discovered much earlier by Bouguer) relates the incident and transmitted intensities of a beam of light passing through an aerosol to the length, l, of the aerosol:

$$I = I_0 \exp(-Kl) \qquad (4\text{-}15)$$

K is commonly called the extinction coefficient of the aerosol, Kl is its turbidity, and I/I_0 is the transmittance. Absorbance is defined as $\log(I_0/I)$. It is directly proportional to the concentration of particles in an aerosol and is the quantity often indicated by instruments designed to determine concentrations by light extinction.

The effectiveness of an individual particle to produce extinction can be described by the particle extinction coefficient:

$$E \equiv \frac{\text{flux absorbed and scattered by the particle}}{\text{flux incident on the particle}} \qquad (4\text{-}16)$$

E is also known as the area efficiency factor. The Lambert-Beer law can be written in terms of E:

$$\frac{I}{I_0} = \exp(-NaEl) \qquad (4\text{-}17)$$

where N is the number concentration of monodispersed particles of projected area a. When the particles are polydispersed, NaE must be integrated over the particle size distribution. However, if E is constant, a single value of a can be used, namely the area of the particle of mean surface diameter, $\bar{d}_{20}$ (see Table 1-3). Hodinkson[6] has pointed out that if the particles are not too reentrant, and are oriented at random, Na is approximately one-fourth of their surface area concentration.

The E for nonabsorbing particles, like Q_{scat}, approaches two as α becomes large. Since the diffracted light is forward-scattered in a direction almost parallel to the incident beam, a light-sensing device, such as a photocell in extinction-measuring equipment, will often receive this light, and the apparent value of E will be unity.

On the other hand, very small nonabsorbing particles, for which α is in the Rayleigh scattering region, scatter equally in the forward and backward directions, and the Lambert-Beer law can be written in the form

$$\frac{I}{I_0} = \exp\left[-\frac{128\pi^5 r^6}{3\lambda^4}\left(\frac{m^2-1}{m^2+2}\right)^2\right]Nl \qquad (4\text{-}18)$$

2.2. Extinction Measurement

Figure 4-23 shows the essentials of extinction-measuring apparatus designed to measure essentially all of the unscattered light and almost none of the scattered light.[6] The first pin hole produces nearly point-source illumination and, together with the collimator lens, produces nearly parallel rays through the aerosol. If these rays are precisely parallel, the only scattered light reaching the photocell is that scattered through

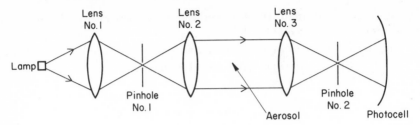

Figure 4-23.   Hodkinson system for measuring extinction.[6]

angles less than the angle determined by the radius of the pin hole and the pole of lens no. 3.

Gumprecht and Sliepcevich[74] used an optical system similar to that shown in Figure 4-23 except for the omission of pin hole no. 1. They proposed a method for correcting for the forward-scattered light received by the phototube, defining a correction factor

$$R = \frac{E_a}{E} \tag{4-19}$$

E_a was defined on the basis of the amount of light scattered by each particle in all directions except within the cone of half-angle ψ which the light-measuring instrument subtends with the particle. Then equation (4-17) becomes

$$\frac{I}{I_0} = \exp\left(-NaREl\right) \tag{4-20}$$

When the angle ψ is less than about 1.4°,

$$R = \frac{1 + [J_0(\alpha\psi)]^2 + [J_1(\alpha\psi)]^2}{2} \tag{4-21}$$

where J_0 and J_1 are Bessel functions of order zero and one, and ψ is expressed in radians.

2.3.   Instrumentation

An important application of extinction measurements is the determination of atmospheric turbidity, which has been defined as any condition of the atmosphere that reduces its transparency to radiation, especially to visible radiation. The sun is usually used as the light source, and a host of instruments of varying degree of complexity have been used.[75-78]

When the sun is used, this is obviously a remote, passive measurement. The most commonly employed instruments are the pyrheliomter and the Volz sun photometer. The former is a differential bolometer or thermo-pile, which is evacuated and protected from convective and conductive heat losses. It is provided with a black surface to absorb solar energy, and a reflecting surface. The differential heating of the surfaces provides a measure of the total flux from the direct sunlight plus the scattered sunlight falling on the blackened surface.

The Volz sun photometer is inexpensive, simple to use, and is the basis for a network of turbidity-measuring stations in the United States.[78] It is shaped like a small box, 5.5 × 2.2 × 2.2 in., with a diopter on the side that measures the solar elevation angle and is calibrated in terms of the optical air mass. An aperture plate is placed at one end of the box that provides a viewing angle of 3.75°. Light from the sun passes through a Kodak Wrattan 65 filter and illuminates a selenium photocell, the output of which is indicated by a micrometer in the top of the box. These photometers are calibrated against standard photometers. The standards are Volz photometers that use IP28 phototubes which respond linearly, instead of selenium photocells which have a nonlinear response. An 11-wavelength photometer has also been developed.

The results of turbidity measurements are usually reported as turbidity coefficients (B), which have been defined in several ways. Volz has adopted the definition of Schüepp[79] for use with his photometer:

$$\log \frac{J}{J_0} = -(\tau_r + \tau_x + B) m \qquad (4\text{-}22)$$

where J is the observed solar radiation of wavelength 0.5 μm, adjusted to the mean sun-earth distance; J_0 is the solar radiation at this wavelength outside the atmosphere at the mean sun-earth distance; τ_r and τ_x are the scattering coefficients for air molecules and the absorption coefficient for ozone, respectively, and m is the optical air mass. Linke and Boda,[80] and Ångström[77,81] have provided other methods for expressing turbidity.

A difficulty in estimating the particulate loading of the atmosphere by measuring solar radiation is that undetected tenuous cirrus contributes to the turbidity. Robinson[75] has suggested that such cirrus often occurs.

Transmission measurements are often made in ducts and stacks as indicators of the concentrations of particulate material in the gases flowing through them. The instruments usually consist of little more than a lamp, a couple of lenses, and a photocell. Therefore, except for particles scattering in the Rayleigh region, considerable scattered light reaches the photocell. Numerous devices for making transmission measurements in ducts and stacks are commercially available.

Various transmissometers have been developed for measuring the extinction coefficients of very dilute aerosols, such as photochemical smog. One used by the author at Stanford Research Institute was based on that by Bradbury and Fryer.[82] A mechanically chopped light beam was reflected back and forth a number of times to produce a long path length and was then focused on a photocell.

Steffens[83] developed a photographic method for determining extinction coefficients (K) in the open atmosphere. A black object silhouetted against the sky near the horizon is photographed or, if for some reason the sky near the horizon cannot be used, two black objects at different distances are photographed simultaneously (Figure 4-24). The optical density, D, of a photographic negative is given by the relationship

$$D = g + \gamma \log (\text{exposure}) \tag{4-23}$$

where g and γ are constants over a wide range of densities. The exposure is $Ixf(\text{time})$ where $f(\text{time})$ reduces to time if the reciprocity law for photographic materials applies. Then the ratios of the intensities defined by Figure 4-24 can be calculated from the equation

$$\frac{I_2}{I_1} = \log^{-1} \frac{D_2 - D_1}{\gamma} \tag{4-24}$$

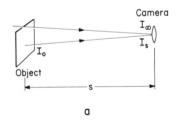

a

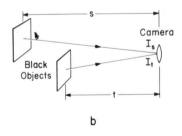

b

Figure 4-24. Arrangements for determining the extinction coefficient. Reprinted with permission from Steffens, C., *Ind. Eng. Chem.* **41**, 2396 (1949). Copyright by the American Chemical Society.

where D_2 and D_1 are the densities of the images, and γ is calculated from other measurements of density on the negative. The extinction coefficient can then be calculated from

$$K = \frac{1}{s} \ln \frac{1}{1 - (I_s/I_\infty)} \qquad (4\text{-}25)$$

or

$$\frac{I_s}{I_t} = \frac{1 - e^{-Ks}}{1 - e^{-Kt}} \qquad (4\text{-}26)$$

Equation (4-26) cannot be solved explicitly for K, but a table can be constructed relating K to I_s/I_t for given values of s and t. Of course, methods other than photographic can be used to obtain I_s and I_t or I_∞.

Steffens technique was developed primarily to estimate visual range, L, defined in Chapter 3. Assuming a contrast limen ϵ of 0.02,

$$L = \frac{1}{K} \ln \frac{1}{\epsilon} = \frac{3.912}{K} \qquad (4\text{-}27)$$

which is essentially the same as equation (4-12). Steffens used the following approach to estimate size distributions from the extinction coefficients. He assumed a size distribution of the form

$$f(d) = ad^{-b} \qquad (4\text{-}28)$$

where the exponent b is to be determined from the extinction coefficients. It is readily shown that

$$K = \pi a E \left(\frac{\lambda}{2\pi}\right)^{3-b} \qquad (4\text{-}29)$$

where E can be calculated from Mie theory. Extinction coefficients are measured for two wavelengths (λ_1 and λ_2). Then,

$$b = 3 - \frac{\log (K_2/K_1)}{\log (\lambda_2 - \lambda_1)} \qquad (4\text{-}30)$$

This approach was used by Steffens and Rubin[84] to determine b for Los Angeles smog. Extinction coefficients were determined by photographing the black targets on panchromatic film through a series of filters such as Wratten A (No. 25), B (No. 28), and C5 (No. 47); red, green, and blue, having effective wavelengths of 0.62, 0.54, and 0.45 μm, respectively. They obtained an arithmetic mean value for b of 4.5.

Knollenberg[85] has used an electro-optical technique for measuring the size distributions of cloud and precipitation particles. An array of photo-detectors forms a size-measuring grid in a shadowgraph-type of imaging system. A collimated light beam illuminates the grid, and each particle passing through the beam casts a shadow which is geometrically related to the particle diameter. An array of light-transmitting fibers is packed in a row on one end to form the optical grid and is spread out to individual photomultiplier tubes on the other end.

A very different approach from any of those described so far to detect and size particles has been used by Schuster and Knollenberg,[86] Proctor,[87] and Schleusener and Read.[88] It is based on the fact that particles injected into the radiation field of a laser cavity create losses due to scattering, resulting in a power loss much greater than that due to the geometric loss of the usual extinction measurements. Schuster and Knollenberg used a He-Ne laser. They pointed out that the cavity loss can be computed from a knowledge of the transverse mode structure of the laser and the optical total cross section of the aerosol particle. When the particle radii exceed 3λ, the effective cross section is twice the geometric cross section, as discussed earlier. For smaller particles, the interferometric behavior of the cavity suppresses the normal Mie resonances by converting them to very narrow spikes.

Searchlights have long been used to investigate airborne particulate material. Some of the most recent work has been done by Elterman et al.[89,90] Their method involved the use of a modulated searchlight beam projected upward at a 75° elevation. The scattered light intensity along the beam was measured about thirty km from the searchlight with a parabolic light collector and photomultiplier. The data yield vertical profiles of the aerosol extinction coefficient and thus information concerning the aerosol layer structure and its properties. Rayleigh scattering by air molecules was calculated by using standard tables for atmospheric properties as a function of altitude, and the results reported as turbidity, defined as the ratio of aerosol to Rayleigh extinction coefficients. Measurements were made for altitudes up to 25 km.

REFERENCES

1. Kerker, M., *The Scattering of Light and Other Electromagnetic Radiation*, Academic Press, New York, 1969.
2. Mie, G., *Ann. Physik*, **25**, 277 (1908).
3. Lorenz, L., *Videnskab. Selskab. Skrifter*, **6**, (1890).

4. Rayleigh, Lord (J. W. Strutt), *Phil. Mag.,* **41,** 107, 274, 447 (1871).

5. van de Hulst, H. C., *Light Scattering by Small Particles,* Wiley, New York, 1957.

6. Hodkinson, J. R., in *Aerosol Science,* C. N. Davies, Ed., Academic Press, London, 1966.

7. Sinclair, D., in *Handbook on Aerosols,* Atomic Energy Commission, Washington, D. C., 1950.

8. Tabibian, R. M., W. Heller, and J. N. Epel, *J. Coll. Sci.,* **11,** 195 (1956).

9. Tabibian, R. M., and W. Heller, *J. Coll. Sci.,* **13,** 6 (1958).

10. Gucker, F. T., and R. L. Rowell, in *The Physical Chemistry of Aerosols,* General Discussion No. 30, The Faraday Society, Aberdeen, 1961.

11. Gucker, F. T., and J. J. Egan, *J. Coll. Sci.,* **16,** 681 (1961).

12. Phillips, D. T., P. J. Wyatt, and R. M. Berkman, *J. Coll. Interface Sci.,* **34,** 159 (1970).

13. Farone, W. A., and M. Kerker, *J. Opt. Soc. Am.,* **56,** 481 (1966).

14. Cooke, D., and M. Kerker, *Rev. Sci. Inst.,* **39,** 329 (1968).

15. Cooke, D., and M. Kerker, *J. Opt. Soc. Am.,* **59,** 43 (1969).

16. Lundberg, R. K., *J. Coll. Interface Sci.,* **29,** 565 (1969).

17. Hodkinson, J. R., in *Procedings of the Interdisciplinary Congress on Electromagnetic Scattering,* M. Kerker, Ed., Pergamon Press, New York, 87 (1963).

18. Aden, A. L., and M. Kerker, *J. App. Phys.,* **22,** 1242 (1951).

19. Sinclair, D., and V. K. LaMer, *Chem. Revs.,* **44,** 245 (1949).

20. Kerker, M., and V. K. LaMer, *J. Am. Chem. Soc.,* **72,** 3516 (1950).

21. Maron, S. H., and M. E. Elder, *J. Colloid Sci.,* **18,** 199 (1963).

22. Maron, S. H., P. E. Pierce, and M. E. Elder, *J. Colloid Sci.,* **19,** 591 (1964).

23. Butcher, S. S., and R. J. Charlson, *An Introduction to Air Chemistry,* Academic Press, New York, 1972.

24. Beuttell, R. G., and A. W. Brewer, *J. Sci. Instrum.,* **26,** 357 (1949).

25. Ahlquist, N. C., and R. J. Charlson, *J. Air Poll. Contr. Assoc.,* **17,** 467 (1967).

26. Ahlquist, N. C., and R. J. Charlson, *Environ. Sci. Technol.,* **2,** 363 (1968).

27. Ahlquist, N. C., and R. J. Charlson, *Atmos. Environ.,* **3,** 551 (1969).

28. Charlson, R. J., H. Horvath, and R. F. Pueschel, *Atmos. Environ.,* **1,** 469 (1967).

29. Charlson, R. J., N. C. Ahlquist, and H. Horvath, *Atmos. Environ.,* **2,** 455 (1968).

30. Thielke, J. F., R. J. Charlson, J. W. Winter, N. C. Ahlquist, K. T. Whitby, R. B. Husar, and B. Y. H. Liu, *J. Colloid Interface Sci.,* **39,** 252 (1972).

31. Chaney, A. L., Direct photography of aerosol suspensions, in *Air Pollution,* Louis McCabe, Ed., McGraw-Hill, New York, 1952.

32. LaMer, V. K., and P. K. Lee, A forward angle scattering camera which measures polydispersity in the range 0.1 to 0.5 microns, in *Report of Symposium V Aerosols*, Chemical Corps, Army Chemical Center, Maryland, 1953; Rev. Sci. Inst., **25**, 1004 (1954).

33. Gucker, F. T., Jr., C. T. O'Konski, H. B. Pickard, and J. N. Pitts, Jr., *J. Am. Chem. Soc.,* **69**, 2422 (1947).

34. Gucker, F. T., Jr., and C. T. O'Konski, *Chem. Rev.,* **44**, 373 (1949).

35. Gucker, F. T., Jr., and C. T. O'Konski, *J. Colloid Sci.,* **4**, 541 (1949).

36. Fisher, M. A., S. Katz, and A. Lieberman, in *Proceedings of the Third National Air Pollution Symposium,* Stanford Research Institute, Menlo Park, Calif., 1955.

37. Zinky, W. R., *J. Air Poll. Contr. Assoc.,* **12**, 578 (1962).

38. Brown, P. M., *Environ. Sci. & Tech.,* **3**, 768 (1969).

39. Rosen, J. M., *J. Geophys. Res.,* **69**, 4673 (1964).

40. Rosen, J. M., *Space Sci. Rev.,* **9**, 58 (1969).

41. Rosen, J. M., *J. Geophys. Res.,* **73**, 479 (1968).

42. Rosen, J. M., in *Climatic Impact Assessment Program Monograph No. 1,* U.S. Dept. of Transportation, Washington, D.C., in press.

43. Hodkinson, J. R., and J. R. Greenfield, *Appl. Optics* **4**, 1463 (1965).

44. Quenzel, H., *Appl. Optics,* **8**, 169 (1969).

45. Quenzel, H., *Appl. Optics.,* **9**, 1931 (1970).

46. Whitby, K. T., and R. A. Vomela, *Environ. Sci. Tech.,* **1**, 801 (1967).

47. Whitby, K. T., B.Y.H. Liu, and R. A. Vomela, *Evaluation of Optical Particle Counters,* Final Report AP00283-1, Mechanical Engineering Dept., U. of Minnesota, Minneapolis Minn., June, 1967.

48. Thomas, A. C., A. N. Bird, R. H. Collins, and P. C. Rice, A portable photoelectric aerosol counter and particle size analyzer, Inst. Soc. of Am. Preprint No. 23-SF60, 1960.

49. Whitby, K. T., D. A. Lundgren, and C. M. Peterson, *Int. J. Air Water Pollution,* **9**, 263 (1965).

50. Rimberg, D., and D. Keafer, *J. Colloid Interface Sci.,* **33**, 628 (1970).

51. Cadle, R. D., and E. J. Wiggins, A.M.A. *Arch. Ind. Health,* **12**, 584 (1955).

52. Cannon, T. W., *J. Appl. Meteor.,* **9**, 104 (1970).

53. Cannon, T. W., *Image Technology,* April/May, 1970.

54. Stroke, G. W., *An Introduction to Coherent Optics and Holography,* 2nd ed., Academic Press, New York, 1969.

55. Koechner, W., *Industrial Res.,* 44-48, April, 1973.

56. Derr, V. E., ed., *Remote Sensing of the Troposphere,* U.S. Dept. of Commerce, National Oceanic and Atmospheric Administration, Washington, D.C., 1972.

57. Collis, R. T. H., *Appl. Optics,* **9**, 1782 (1970).

58. Fox, R. J., G. W. Grams, B. G. Schuster, and J. A. Weinman, *J. Geophys. Res.*, **78**, 7789 (1973); Grams, G. W., and C. M. Wyman, *J. Appl. Meteor.*, **11**, 1108 (1972).

59. Hiltner, W. A., *Science,* **109**, 165 (1949); *Nature,* **163**, 283 (1949).

60. Hall, J. S., *Science,* **109**, 166 (1949).

61. Hiltner, W. A., *Astrophys. J.,* **109**, 471 (1949).

62. Hall, J. S., and A. H. Mikesell, *Astron. J.,* **54**, 187 (1949).

63. Davis, L., and J. L. Greenstein, *Astrophys. J.,* **114**, 206 (1951).

64. Spitzer, L., Jr., and J. W. Tukey, *Astrophys. J.,* **114**, 187 (1951).

65. Lyot, B., *Ann. Obs. Paris-Mendon*, **8**, 70 (1929).

66. Hansen, J. E., and J. V. Hovenier, Nature of the Venus Clouds as Derived from their Polarization. Presented at Copernicus Symposium IV, Exploration of the Planetary System, Torum, Poland, September 5–8, 1973. To be published in *Proceedings of IAU Symposium 65,* Reidel, Dordrecht, Holland; *J. Atmos. Sci.,* **31**, 1137 (1974).

67. Rose, A. H., J. S. Nader, and P. A. Drinker, *J. Air Poll. Contr. Assoc.,* **8**, 112 (1958).

68. Dave, J. V., and C. L. Mateer, *J. Geophys. Res.,* **73**, 6897 (1968).

69. Kerker, M., E. Matijevic, W. Espenscheid, W. Farone, and S. Kitani, *J. Colloid Sci.,* **19**, 213 (1964).

70. Kerker, M., A. L. Cox, and M. D. Schoenberg, *J. Colloid Sci.,* **10**, 413 (1955).

71. Chin, J. H., C. M. Sliepcevich, and M. Tribus, *J. Phys. Chem.,* **59**, 845 (1955).

72. Kratokvil, J. P., and C. Smart, *J. Collid. Sci.,* **20**, 875 (1965).

73. Wilson, L. G., and P. Cavanagh, *Atmos. Environ.,* **3**, 47 (1969).

74. Gumprecht, R. O., and C. M. Sliepcevich, *J. Phys. Chem.,* **57**, 95 (1953).

75. Robinson, G. D., in *Man's Impact on the Climate,* W. H. Matthews, W. W. Kellogg, and G. D. Robinson, Eds., The MIT Press, Cambridge, Mass., 1971.

76. Kondratyev, K. Y., *Radiation in the Atmosphere,* Academic Press, New York, 1969.

77. Drummond, A. J., Ed., *Precision Radiometry,* Advances in Geophysics *14,* Academic Press, New York, 1970.

78. Flowers, E. C., R. A. McCormick, and K. R. Kurfis, *J. Appl. Meteor.,* **8**, 955 (1969).

79. Schüepp, W., *Meteor. Geophys. Bioklim.,* **B1**, 257 (1949).

80. Linke, F., and K. Boda, *Meteor. Z.,* **39**, 161 (1922).

81. Ångström, A., *Geograph. Ann.,* **11**, 156 (1929).

82. Bradbury, M. E., and E. M. Fryer, *Bull. Am. Meteor. Soc.,* **21**, 391 (1940).

83. Steffens, C., *Ind. Eng. Chem.,* **41**, 2396 (1949).

84. Steffens, C., and S. Rubin, in *Proceedings of the First National Air Pollution Symposium,* Stanford Research Institute, Menlo Park, Calif., 1949.

85. Knollenberg, R. G., *J. Appl. Meteor.,* **9,** 86 (1970).

86. Schuster, B. G., and R. Knollenberg, *Appl. Optics.,* **11,** 1515 (1972).

87. Proctor, T. D., *J. Sci. Instrum.,* **1,** 631 (1968).

88. Schleusener, S. A., *J. Air Poll. Cont. Assoc.,* **19,** 40 (1969).

89. Elterman, L., R. B. Toolin, and J. D. Essex, *Appl. Optics,* **12,** 330 (1973).

90. Elterman, L., *Appl. Optics,* **5,** 1769 (1966).

5

MULTISTAGE IMPACTORS AND CENTRIFUGAL CLASSIFIERS

1. IMPACTORS

The theory and application of single-stage impactors was described in Chapter 2. The emphasis was on the collection of particulate material for subsequent analysis. However, impactors can also be used as particle-sizing instruments by incorporating several of them, each having a different lower size limit for collection, into a single device. The theory of these devices is, of course, that for the single-stage instruments. The individual impactors can be operated either in series or in parallel.

One of the simplest multistage impactors is merely an extension of the moving rod or fiber impactor shown schematically in Figure 2-26. Fibers or rods of various diameters can be mounted on the same rotating frame, or each on a different frame. Since the cut-off size for particle collection decreases with decreasing fiber diameter for a given velocity relative to the aerosol, measurements of the quantities of particulate material of each of the several fibers or rods can be used to estimate particle size distribution. The measurements can be made by weighing, by chemical analysis, or by microscopy. A commercially available instrument that can be used in this manner is the Rotorod Sampler, manufactured by Metronics Associates. Since it is powered with a 12-V dry cell and weighs only 6 oz it can be used almost anywhere. Modifications of this approach have been used to determine size distributions of supercooled cloud droplets from aircraft. One modification involved the use of rotating cylinders of differing diameter mounted coaxially above the aircraft. The collected droplets froze on the cylinders. By weighing the amount collected on each cylinder, and assuming the general form of the distribution curve, particle size distributions and particle concentrations could be estimated.[1,2]

The aircraft impactors shown in Figures 2-27 and 2-28 can also be modified to be multistage impactors by constructing each impactor surface with stepwise decreasing widths.

A much better approach is to use jet impactors mounted in series. The orifice of each stage is progressively smaller, and thus the linear speed through each stage is progressively greater. The particles are classified into overlapping size fractions. The aerosol is sucked through the instrument with a vacuum pump, and if the pressure drop across the smallest orifice exceeds one-half atmosphere, the linear air speed through the orifice is essentially sonic. The term cascade impactor was coined by May,[3] who seems to have been the first to construct such a device.

One of the best known of these instruments is the Casella cascade impactor, shown in Figure 5-1. It was designed to collect particles in the size range 200–0.5 μm diameter. Four jets are arranged at right angles to one another, and the impactor is designed to operate with a pump capable of pumping 17.5 liters of air per minute at about 0.1 atmospheric pressure. The collecting surfaces are glass disks that are usually coated with an adhesive to help retain the particles, unless the particles are droplets.

The main disadvantages of this impactor are that it is awkward to assemble after the glass disks are put in place, and that it has a tendency to come apart at critical times.

A much simpler impactor, that is easier to use, is the Unico cascade impactor, manufactured by Unico Environmental Instruments, Inc. Figure 5-2 is a photograph of this instrument which has been modified slightly to simplify its operation, so that it can accept microscope slides on which are mounted electron microscope grids. It was designed by Lippman[4] specifically for large-scale field sampling programs; it is lightweight, small, easily loaded and unloaded, and easily cleaned. The particulate material is collected at four impaction stages on two standard microscope slides with two deposits on each slide.

A very useful feature is the manual slide movement mechanism. Essentially, it is a metal plate with four arms, hinged to the plate, to hold the slides in place and to guide their movement. They are spring-loaded and are shown in the closed position in the figure. The slides can be advanced to permit taking a large number of samples. Movement in the

Figure 5-1.   Casella cascade impactor. a) Assembled; b) Partially disassembled, showing one of the jets.

unmodified version is in 1/16-in. increments, permitting the collection of nine samples. The modified version shown in the figure is arranged to move the slides to any desired position with the rack and pinion shown at the upper right. This provides for centering electron microscope grids mounted on the slides opposite the narrowest jet. Another modification is a slight depression in the impactor surface just to the right of the narrowest jet to accommodate the electron microscope grids.

Figure 5-2.   Modified Unico cascade impactor.

The last stage is followed by a filter holder to collect any particles not retained on the microscope slides. Lippman suggested the use of Dow Corning 200 fluid on the slides, and cleaning the impactor with benzol or one of its homologs. Lippman prepared a nomogram for use with his instrument (Figure 5-3).

Wall losses for this impactor were investigated and found not to exceed 30%; they were believed not to materially affect its use for determining particle size distributions.

A very different arrangement, designed to sample at a much higher flow rate than the Casella and Unico instruments, is pictured in Figures 5-4 and 5-5. It is designed to attach to a standard high volume sampler

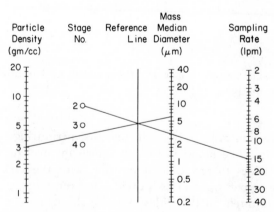

Figure 5-3.   Nomogram of Unico impactor.[4] By permission of the American Industrial Hygiene Association.

Figure 5-4. High volume cascade impactor. Courtesy of Sierra Instruments, St. Paul, Minn.

Figure 5-5. Disassembled high volume cascade impactor. Courtesy of Tech-Ecology, Inc., Mountain View, Calif.

and to classify the suspended particles into up to six fractions. Nominally, for particles having a density of one g cm^{-3}, these fractions are 10 μm and greater diameter, 4.9 to 10 μm, 2.7 to 4.9 μm, 1.3 to 2.7 μm, 0.61 to 1.3 μm, and the remaining particles collected on the back-up filter of the high volume sampler. The flow rate range is 20–60 ft^3 min^{-1}. Four models are available, having 5, 4, 3, or 1 impactor stages; slotted collection paper is also available.

The aerosol enters the impactor through the parallel slots in the first stage, and particles larger than the cut-off size are collected by impaction on the slotted collection paper. The aerosol then passes through the slots in the collection paper, on to the second impactor stage, and so forth. The collection papers and back-up filter are weighed before and after sampling to obtain the particle size distribution.

The greatest advantage to this approach is the high sampling rate, whereas the chief disadvantages are that the particles are collected on a surface that makes examination by microscopy inconvenient, and that an appropriate high-volume sampler must be provided.

Sierra Instruments also sells a source cascade impactor. It is designed to be used within a stack or other source of particulate pollution, and consists of six impactor stages having particle size cut-offs ranging from 0.5 to 15 μm diameter, and a final filter holder. It is constructed of stainless steel, and the paper on which the samples are impacted can withstand temperatures as high as 500°F. It easily fits through a 3-in. i.d. port. It requires the use of special slotted collection paper. Each stage is essentially a disk with slots extending radially from the center, and the manufacturer claims that this design results in very low wall losses. A larger version, called the Model 215 ambient cascade impactor, is also available.

The Anderson sampler was one of the first if not the first impactors with a number of jets at each stage to increase the rate of collection of particulate material. The initial model[5] was designed to collect airborne microorganisms on petri dishes, and a later one[6] was designed primarily for general respiratory health-hazard assessment. The six-stage standard sampler is 3.75 in. in diameter, 6 in. high, and weighs 2.5 lb. Air leakage is prevented with 0-ring seals. Each stage consists essentially of two circular plates. Four hundred jets are drilled in the upper one, and the lower, collection plate of glass or stainless steel is positioned 2.5 mm below. The air-sampling rate is about one cfm. Several variations have been produced, such as the Andersen mini-sampler and the Andersen stack sampler.

Ludwig[7] prepared a simple numerical analogue to cascade impactors and applied it to the standard Andersen sampler. Ideally, each stage of an impactor collects all particles larger than a certain size and none smaller. However, as discussed in Chapter 2, the collection efficiency varies from 0 to 100% monotonically over a range of diameters. The diameter chosen to represent the "cut-off" point is usually that for which the collection efficiency is 50%, as suggested by Mercer,[8] who also suggested that the real collection efficiency curves might be approximated with a sloping straight line. If the amounts of uncollected larger particles and collected smaller particles are equal, each real impactor stage will collect the same amount of material as the ideal stage. The diameter that yields this equality is called the effective cut-off diameter of that stage.

As Ludwig points out, the effective cut-off diameter is a function of the size distribution of the particles sampled as well as of the characteristics of the impactor stage. For instance, this diameter is decreased by increasing the number of smaller particles and decreasing the number of larger particles in the region of the stage's cut-off. Also, the effective cut-off diameter for a particle number distribution differs from that for a particle mass distribution for the same sample.

The following general equation was used as the basis for the model:

$$G_m = \int_0^\infty E_m(d) \, f_m(d) \, d(d) \qquad (5\text{-}1)$$

where $E_m(d)$ is the collection efficiency of the mth stage as a function of particle diameter and $f_m(d)$ is the size distribution function of the material entering that stage. The term $f_m(d)$ can be a conventional distribution function, in which case G_m is the total number of particles on stage m; or $f_m(d)$ can be a mass distribution function, in which case G_m will be the mass of material collected on the mth stage.

The Andersen sampler was simulated numerically using empirical collection efficiencies for each stage. The numerical analog was used to sample particle populations log-normally distributed according to equation (1-43). However, almost any size distribution that can be expressed in mathematical form can be sampled with the model. Ludwig reached the following conclusions:

1. Usual data reduction methods overestimate the geometric standard deviation, s_g, of the particle size distributions.

2. The error is small for values of s_g greater than about 4, but becomes significant for values less than 2.

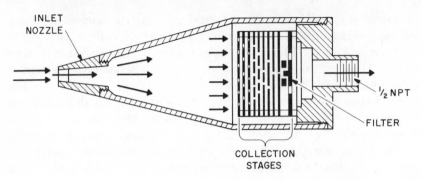

Figure 5-6. Environmental Research Corporation TAG impactor.

3. The magnitudes of the errors vary with the shape of the particle size distribution over the size range sampled by the impactor.

The Environmental Research Corporation manufactures the TAG impactor, shown schematically in Figure 5-6, designed for field or laboratory use. The cone shape of the inlet housing is intended to decelerate the aerosol sample and minimize particle loss to the walls. The cut-off diameter of the first of the nine stages is 30 μm for particles of unit density, and that of the last of the stages is 0.6 μm. A tenth stage consists of an "absolute" fiber filter. The nominal air flow rate is 0.5 cfm, and the collector surfaces are stainless steel. The overall dimensions are 3 in. o.d. by 9 in.

Lundgren[9] has designed a four-stage impactor that collects the particulate matter on four rotating drums, each with a collecting surface area of about 10 in². It is especially useful for long sampling times, such as 24 hours, and for sampling very dusty atmospheres. The large collection surface allows a large quantity of particulate matter to be collected with little danger of particle build-up and re-entrainment. Figure 5-7 shows the arrangement of jets and cylinders.

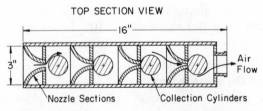

Figure 5-7. Lundgren impactor. Reprinted with permission from *J. Air Pollution Control Assoc.* **17**, 225 (1967).

Lundgren also compared the wall losses for his impactor with those for the Unico, Casella, and Andersen impactors operating at their normal flow rates of 3.0, 0.61, 0.61, and 1.0 cfm, respectively. Aerosols of uranine dye particles were sampled simultaneously and isokinetically (defined in Chapter VII) with an impactor and with a filter. Each impactor inlet, each impaction surface, each nozzle section, and the filter was washed separately, and the fluorescence of the solutions measured. The results are shown in Figure 5-8. Lundgren pointed out that although wall loss cannot be eliminated completely, it can be determined and corrected for.

Gordon et al.[10] also compared the Lundgren impactor with others, namely the Andersen and an impactor designed by the Battelle Memorial Institute (manufactured by Scientific Advances, Inc.) which has a single jet at each stage. They analyzed collections from air for various elements on each stage: V, Ca, Al, Br, Mn, Na, Cl. All three impactors were "reasonably good," but all performed worst for calcium and aluminum, presumably from dirt and fly ash. Such particles are hard and dry, and tend to bounce off the collecting surfaces. The Battelle instrument had the best overall collection efficiency. Failure to retain particles on the collection surfaces ("bounce-off") was especially evident for the Lundgren instrument, but it was markedly improved by using sticky surfaces.

GCA/Technology Division manufactures the respirable dust monitor shown in Figure 5-9. It is designed for field measurements of the respirable fraction or the total mass loading of dust particles. The monitor is fully self-contained and provides automatic and direct digital readout

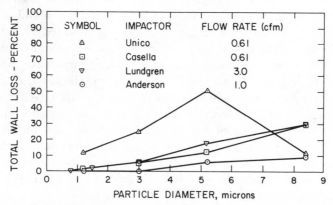

Figure 5-8. Total wall loss versus particle size for different impactors. Reprinted with permission from *J. Air Pollution Control Assoc.* **17**, 225 (1967).

Figure 5-9.   Respirable dust monitor. Courtesy of GCA/Technology Division.

of the mass concentration of airborne dust. Two stages are employed. The first is a cyclone precollector which retains the nonrespirable fraction of the particles. It is designed to retain essentially all particles larger than about 10 μm in diameter for spheres of unit density, and to pass all particles smaller than about 2 μm. Particles not collected by the cyclone are collected by the second stage, which consists of a circular nozzle impactor and a beta-radiation absorption system. Particulate material is collected by impaction on a thin plastic film and absorbs beta radiation from a carbon-14 source which otherwise would reach a Geiger counter. The absorption of low-energy beta radiation depends almost entirely on the mass per unit area of the absorbing material and the maximum energy of the impinging electrons. It is nearly independent of the chemical composition or the physical characteristics of the absorbing matter. The readout is in milligrams of dust per cubic meter of air. The instrument is powered by Ni-Cd rechargeable batteries.

An unusual method for determining particle concentrations and size distributions involves the use of a single-stage, variable-slit impactor

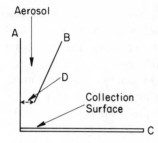

Figure 5-10. Geometry of the variable-slit impactor.

combined with a particle counter that counts (but does not size) the particles escaping the impactor.[11] The arrangement of the impactor is shown in Figure 5-10. The impactor slit walls are labeled A and B. A is fixed, but B can be moved along its plane; when B is moved downward the slit width, D, decreases. The movement of B is controlled with a micrometer, and the particles are impacted onto a glass slide held in place with a clip. The particle counter used to measure the numbers of particles escaping the impactor was a Royco Model 245 Aerosol Monitor.

Because of the inherent scaling features resulting from the final geometry. this impactor system can be calibrated using a single monodispersed aerosol at one flow rate merely by varying the slit width. Thus the calibration is much simpler than that for the cascade impactor, which requires a separate calibration for each stage. Furthermore, the variable-slit impactor with particle counter has a more rapid response and more rapid output of size distribution data than have cascade impactors. A disadvantage of the variable slit impactor is that it must either be automated or manually operated to scan through a range of slit widths, whereas the cascade impactors have no moving parts and can be operated unattended for long time periods.

May[12] has described a three-stage liquid impinger which operates in much the manner of the impingers described in Chapter 2. It is designed to sample viable airborne organisms and to determine their concentrations within size ranges defined in terms of regional respiratory tract deposition. It is manufactured by A. W. Dixon and Co., London.

2.   CENTRIFUGAL CLASSIFIERS

Most centrifugal particle classifiers operate by imparting a spiral flow to the aerosol. Particles move through the aerosol toward the outer surface at a rate that is a function of their size, shape, and density, and of the angular velocity of the flowing aerosol. If the particles are all of one shape and density and if the stream of aerosol is shallow, particles of nearly one size will be deposited at each distance along the collecting surface. If the stream is relatively thick, at each distance along the collecting surface the particles deposited will all be smaller than the minimum size which will traverse the entire thickness of the stream in the time required for the aerosol to flow to that position. Laminar rather than turbulent flow is, of course, essential.

One of the earliest and best-known instruments of this type is the "conifuge," developed by Sawyer and Walton.[13] Essentially, it is a conical centrifuge. An inner, solid cone is within a hollow cone of the same angle, 60°, and separated from it by 5.8 mm. The cones are mounted within a nonrotating housing and rotate together at a speed of about 3000 rpm. A tube passes through the housing and the apex of the outer cone. Rotation of the cones draws air through the tube, causing the air to spread out over the apex of the inner cone. The separation between the cones is small relative to the mean radii of the cones and the distances the aerosol travels, so the particles are nearly completely separated according to "aerodynamic size." The sampling rate is usually about 25 cm^3 min^{-1}.

A small portion of the deposited material is collected for microscopic examination on two narrow glass slides that are T shaped in cross section. Diametrically opposite slots extend nearly the entire length of the outer cone into which the slides are inserted and held so that each is nearly flush with the cone surface. The instrument is effective for sampling spheres of unit density over the range 0.5–30 μm diameter. Larger particles are often lost by impaction on the inner cone. The sampling speed, controlled by an orifice on the sampling tube, can be increased to 75 cm^3 min^{-1} when no particles of unit density larger than 10 μm diameter are present.

Green and Lane[14] suggested the following graphical method for deriving the size distribution from the particles deposited on the slides. The total number of particles, N, deposited on the outer cone is given by the equation

$$N = 2\pi\Sigma N_D R \tag{5-2}$$

where N_D is the number of particles per square millimeter at the distance D from the point of initial collection, and R is the radius of rotation in millimeters at the distance D. The values of N_D are determined with a microscope from the glass slides for a number of values of D. Then $N_D R$ is plotted against D, and N is the area under the graph.

Since each value of D corresponds to a single particle size, replacing the values of D along the D axis with the corresponding particle sizes obtained by a prior calibration converts the curve to a size distribution curve. A similar approach can be used to obtain the mass distribution, using the equation

$$M = \frac{\pi^2 \rho \Sigma N_D R d^3}{3} \tag{5-3}$$

The area under the curve $N_D R d^3$ versus D is proportional to the total mass, and the curve becomes the mass distribution curve when the D axis is in terms of particle sizes.

The conifuge suffers from a small sampling rate and the large surface over which the particles are spread. In order to obtain statistically significant numbers of particles on the glass slides at each distance D, the aerosols sampled must be concentrated or a long sampling time employed. Goetz and his co-workers[15-18] developed a centrifugal aerosol sampler, the "aerosol spectrometer," which largely overcomes these difficulties and will collect particles of much smaller size (>0.03 μm diameter, unit density) than the lower size limit for the conifuge. On the other hand, the upper size limit for the aerosol spectrometer is about 3 μm diameter. Figure 5-11 is a schematic cross section. Like the conifuge, the aerosol spectrometer has a conical rotor, R, but the aerosol sampled is confined to a pair of channels having the form of helical grooves on the outer surface of the rotor (I, II). A detachable cup, B, fits tightly over the rotor, and is lined with a removable foil, F, on which the particles are deposited. The rotor draws air into the system through the inlet, E, and the ports, O, which lead into the channels. The aerosol passes through the channels for two and one-half turns, and, with particles removed, leaves at A. The flow rates through the channels are controlled by orifices, D, below the helical channels. The flow rate through the spectrometer (and thus the residence time) can be varied over a wide range with four sets of orifices or by varying the speed of the rotor. Each set of two orifices, D, is matched by a third orifice inserted at A to maintain a proper balance between the pressure in the helical channels and that in the circular channel in the base. The rotor is driven by a $\frac{1}{4}$ hp motor mounted in the lower part of the housing, G.

The spectrometer tends to heat, so to prevent a change in particle size due to evaporation or thermal decomposition, a refrigeration system is provided that maintains the rotor system temperature within $\pm 0.2°C$. Liquid from a refrigerated unit is pumped through the cavity, K, in the conical insert, L, which is attached to the housing and surrounds the

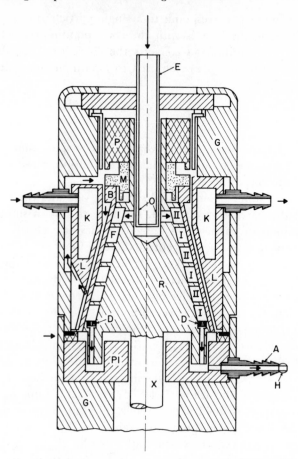

Figure 5-11.   Schematic cross section of the aerosol spectrometer (arrows indicate flow passages). Reprinted by permission from *J. Air Pollution Control Assoc.* **12**, 479 (1962).

rotor at a distance of about 1/16 in. The insert is surrounded by nine vertical channels that communicate with the gap between L and B to facilitate airflow downward along B and then upward through the channels as indicated by the arrows in Figure 5-11. Nine junctions of a sensitive thermopile are mounted in the lower end of L, and the temperature of L is compared with that of the corresponding junctions of the pile that are exposed in a separate unit to a continuous flow of ambient air.

The conical rotor normally spins at about 25,000 rpm, and the sampling rate is 4–8 liters per minute. Unfortunately, the instrument is somewhat subject to orifice losses.

Unlike deposition in the conifuge, deposition in this device is from a rather thick layer of particles, and the deposit along L represents a cumulative distribution along which the number density of particles of a particular size is constant until all have been deposited. Goetz and his co-workers calibrated the aerosol spectrometer with aerosols of polystyrene spheres and developed calibration charts to facilitate the conversion of particle counts along L into particle size distributions.

Goetz and Kallai[18] have described a device, which they call the microanalyzer, for evaluating the deposits by ultramicroscopy or by light scattering. The particles are collected on highly polished chromium foil and are illuminated on a microscope stage with a Leitz Ultropak vertical illuminator. Depending on the position of a sliding mirror, the light scattered by each of the particles is recorded on photographic film or the intensity of the light scattered by all the particles in the optical field is measured by a photocell.

Hochreiner and Brown[19] have described two centrifugal samplers, one cylindrical and one conical, that provide discrete size separation. They are based on an earlier instrument by Berner and Reichelt[20] who introduced a new concept which greatly simplifies the construction and operation of such instruments. The Berner and Reichelt instrument allowed the entire inlet system to rotate, avoiding the need for rotating seals. The aerosol entered the device through a ring slit located near the center of rotation, and the air or gas exhausted through nozzles on an outer diameter of the instrument. In order to limit the radial distance the aerosol particles had to travel, the aerosol as it flowed down the rotor was confined to a thin layer by a second layer of room air. Hochreiner and Brown provided a filtering system for this room air to prevent dust particles in such air contaminating the deposit.

The centrifuges were calibrated with aerosols of polystyrene spheres. Figure 5-12 is a photograph of a deposit produced with the conical centrifuge from an aerosol of 0.557 μm diameter polystyrene spheres. The lowest line is the deposit for single spheres and the others are for aggregates of two, three, and so on spheres. The principal lines are close-packed aggregates, whereas the less dense lines, or "ghost lines," between the second and third principal lines and between the third and fourth principal lines are deposits of chains of three and four particles, respectively. This interpretation was verified by electron microscopy.

Both the cylindrical and conical centrifuges collect particles in the aerodynamic diameter range of 0.1–5 μm. The flow rates are quite slow,

Figure 5-12. Sample of 0.557 μm diameter polystyrene spheres taken with the Hochreiner-Brown centrifuge. Reprinted from Hochreiner, D., and P. M. Brown, *Environ. Sci. Techn.* **3**, 830 (1969). Copyright by the American Chemical Society.

about 12 cm³ min⁻¹ and 26 cm³ min⁻¹ for the cylindrical and conical centrifuges, respectively.

Hochreiner and Brown derived the following equation to describe the particle trajectories and the distance travelled before deposition. It was based on an earlier equation of Stöber and Zessack,[21] but included the Cunningham correction:

$$Z = \frac{9\eta F \ln (b/a)}{2\pi^3 f \rho d^2 (b^2 - a^2)(1 + 2A\lambda/D)} \qquad (5\text{-}4)$$

where Z is the distance in an axial direction from the slit ring to the point of deposition, F is the total flow per unit time, a is the inner radius of the outer cylinder of the centrifuge, b is the radius of the inner cylinder of the centrifuge, and A is the Cunningham correction constant (all in cgs units). The following equation applies to the conical centrifuge:

$$d\rho^{1/2} = \left\{ \frac{27\eta Q h_0}{a\pi^3 f^2 \sin\psi \cos\psi[4(l^3 - l_0^3)\sin\psi] + 3a(l^2 - l_0^2)\cos^2\psi} \right\}^{1/2} - A\lambda\rho^{1/2}$$

(5-5)

where Q is the total gas flow, a is the horizontal dimension of the flow channel, h_0 and l_0 are the intial coordinates of the particle in the centrifuge chamber, l is the distance from the entrance to the point of particle deposition, and ψ is the half-angle of the cone (degrees). The values of a and b for the cylindrical centrifuge are 0.5 to 1.0 cm, respectively. The values of a, h_0, l_0 and ψ for the conical centrifuge are 0.57 cm, 0.57 cm, 2.25 cm, and 30°, respectively.

Hochreiner[22] later developed another cylindrical centrifuge which avoids the need for a separate air-cleaning system. The aerosol is split into two streams, and particles are removed from the larger stream in part of the centrifuge. This clean air forms a carrier stream through which the particles in the superimposed, carrier, stream are deposited.

Stöber and Flachsbart[23] employed a design very different from those described above. The essential part was a spiral duct cut into the disk-shaped rotor of the centrifuge. The aerosol particles were collected along a strip of foil about 180 cm in length, which forms the outer wall of the duct. The aerosol was fed into the rotor at the center of rotation, where the aerosol was entrained into a laminar flow of clean air that had entered through an off-center air inlet and passed through a laminator. The clean-air laminator consisted of five thin brass foils mounted parallel to the vertical walls of the rectangular duct. The aerosol flow rates were up to 3 liters per minute. Best results were obtained for rotor speeds of 3000 rpm, and the collection size range was about 0.08–5 μm diameter. This centrifuge provided nearly complete size separation.

One of the simplest centrifugal analyzers is that designed by Matteson et al.[24] (Figure 5-13). The aerosol enters through the central bore and is split into two streams, one passing through the large inlet channel and the other through the small one. Particles in the aerosol entering through the larger port are removed to produce a cumulative size deposition on half of a strip of foil on the outer wall. This process produces clean air for the aerosol injected through the small orifice. The particles in the latter aerosol are deposited in discrete sizes. This instrument has the advantage of confining the aerosol deposit for monodispersed particles to a narrow streak or dot. Matteson et al. developed a theory to

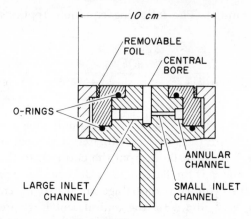

Figure 5-13.   Field-type centrifuge of Matteson *et al.*[24]

describe the relationship between sedimentation distance and the design parameter for this centrifuge.

REFERENCES

1. Diem, M., *Ann. der Hydrogr.,* **70,** 142 (1942).
2. Diem, M., *Met. Rundschau,* **1,** 261 (1948).
3. May, K. R., *J. Sci. Instrum.,* **22,** 187 (1945).
4. Lippmann, S. M., *Ind. Hyg. Assoc. J.,* **22,** 348 (1961).
5. Anderson, A. A., *J. Bacter.,* **76,** 471 (1958).
6. Anderson, A. A., *Am. Ind. Hyg. Assoc. J.,* **27,** 160 (1966).
7. Ludwig, F. L., *Environ. Sci. Technol.,* **2,** 547 (1968).
8. Mercer, T. E., *Ann. Occupational Hyg.,* **6,** 1 (1963).
9. Lundgren, D. A., *J. Air Poll. Cont. Assoc.,* **17,** 225 (1967).
10. Gordon, G. E., E. S. Gladney, J. M. Ondov, T. J. Conry, and W. H. Zoller, *Abstracts,* 167th Annual Meeting, American Chemical Society, Los Angeles, Calif., March 31–April 5, 1974.
11. Cooper, D. W., and L. A. Spielman, *Atmos. Environ.,* **8,** 221 (1974).
12. May, K. R., *Bacter. Rev.,* **30,** 559 (1966).
13. Sawyer, K. F., and W. H. Walton, *J. Sci. Instrum.,* **27,** 272 (1950).
14. Green, H. L., and W. R. Lane, *Particulate Clouds. Dusts, Smokes and Mists,* 2nd ed., Van Nostrand, New York, 1964.
15. Goetz, A., *Geofisica Pura e Applicata,* **36,** 49 (1957).

16. Goetz, A., H. J. R. Stevenson, and O. Preining, *J. Air Poll. Cont. Assoc.*, **10**, 378, 414, 416 (1960).

17. Goetz, A., and O. Preining, in *Physics of Precipitation.* Am. Geophysical Union, Geophysical Monograph No. 5, 164 (1960).

18. Goetz, A., and T. Kallai, *J. Air Poll. Cont. Assoc.*, **12**, 479 (1962).

19. Hochreiner, D., and P. M. Brown, *Environ. Sci. Technol.*, **3**, 830 (1969).

20. Berner, A., and H. Reichalt, *Staub*, **28**, 158 (1968).

21. Stöber, W., and V. Zessack, *Zentralblatt für Biol. Aerosolforschung*, **13**, 263 (1966).

22. Hochreiner, D., *J. Colloid Interface Sci.*, **36**, 191 (1971).

23. Stöber, W., and H. Flachsbart, *Environ. Sci. Technol.*, **3**, 1280 (1969).

24. Matteson, M. J., G. F. Boscoe, and O. Preining, *J. Aerosol Sci.*, **5**, 71 (1974).

6

MISCELLANEOUS METHODS

1. CONDENSATION NUCLEATION

1.1. Theory

One of the more difficult tasks in airborne particle measurement is determining the number concentrations and size distributions of particles with radii <0.1 μm (Aitken particles). One type of instrument that is usually used merely to obtain number concentrations of such particles is the Aitken (condensation) nucleus counter. It is essentially an expansion-type cloud chamber in which water condenses on the particles in a sample of an aerosol. Numerous variations have been designed and applied.

One approach to the theory of heterogeneous nucleation is to consider that the particles act by catalyzing the formation of clusters of water molecules which, in turn, act as nuclei. The condensation nuclei serve to decrease the work of formation of the clusters. The equilibrium vapor pressure of a droplet is slightly greater than that for the same liquid having a plane surface, and is given by the following equation:

$$\ln \frac{p'}{p} = \frac{2M\sigma}{rpRT} \tag{6-1}$$

where M is the molecular weight of the liquid, p and p' are the vapor pressures above a plane and curved surface, respectively, σ is the surface tension, R is the gas constant, and T is the absolute temperature. Equations have also been derived for p'/p when the droplets are charged and when they contain a solute.[1]

In general, the amount of supersaturation required for an insoluble, wettable particle to act as a condensation nucleus will be slightly less than that calculated for a droplet of the same size as the particle. Thus the smaller the particle, the greater the degree of supersaturation required for condensation.

1.2. Instrumentation

The Aitken nucleus counter, in which particles are allowed to settle onto a surface and are counted, is described in Chapter 2. However, many counters used today include a particle counting or measuring system as an integral part of the device. The literature concerning the construction and application of such devices to 1959 was reviewed by Pollak,[2] who himself developed many models of condensation nucleus counters. Most such instruments measure the concentration of the condensed droplets by light attenuation or light scattering. These, of course, are not absolute instruments and must be calibrated. Also, they suffer from the fact that soluble particles will produce different sizes of droplets than insoluble particles. The attenuation measurements are simply made with a beam of light passing through the condensation chamber and a photocell to measure the light extinction. Light scattering is measured with a photomultiplier tube. Commercial versions are available, for example, from the General Electric Co. and Environment One Corp., and operate in an automatic, continuously recording manner with a number of expansion ratios. Rosen[3] has built a condensation nucleus counter using a differential temperature rather than an expansion chamber to obtain supersaturation. Droplets condensed on the nuclei are counted individually by a scattered-light counter, as described in Chapter 4. It was built to fly on balloons in the troposphere and stratosphere.

The Nolan and Pollak[4] version brings an air sample into a moist-wall chamber. The sample is first pressurized with filtered air, and after the air is saturated with water vapor, the pressure is released by opening a valve to the ambient atmosphere. The droplet concentration is measured by light extinction. This instrument has been used to determine particle size distributions by varying the amount of over-pressure and thus the amount of supersaturation. However, a separate calibration must be made for each overpressure.

The Environment One Corp. manufactures a portable, photographic condensation nucleus counter. A supersaturation of at least 300% is provided by the expansion of humidified air. The resulting cloud of droplets is illuminated by a photoflash lamp, and droplet images are recorded on Polaroid film. The number of images per square centimeter of film times a factor defined by the volumetric and optical geometry of the instrument yields the particle concentration in the sampled air. The counts may be made with the naked eye for low-level counts, but a low-power magnifier may be required for particle concentrations in the vicinity of 30,000 particles cm^{-3}.

Rich[5] has developed a variation of the photoelectric nucleus counter specifically to provide information concerning particle size distributions.

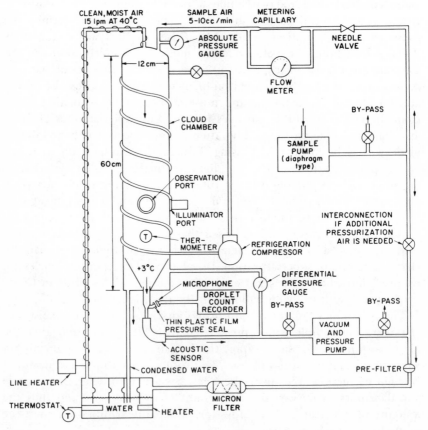

Figure 6.1. Langer condensation nucleus counter.

The sample in the tube of the instrument expands from ambient pressure into two evacuated chambers. The expansion into the first chamber is very rapid, and this determines the supersaturation. The expansion into the second chamber is sufficiently slow that the growth of the water drops limits the supersaturation to below the initial value, but is sufficiently rapid that the expansion is essentially adiabatic. The calibration is essentially the same for all expansion ratios as long as the sum of the two volumes is constant.

Langer[6] has combined a condensation nucleus counter with his acoustic particle counter described later in this chapter. Individual water droplets are counted as they settle through the bottom of the instrument as shown in Figure 6-1.

2. BROWNIAN DIFFUSION

2.1. Theory

When a suspension of particles is in contact with a surface, Brownian motion of the particles having sizes about equal to or smaller than the mean free path of the suspending gas will cause the particles to diffuse to the surface, where they are retained. The theory of such collection by filters was discussed in Chapter 1.

The net transfer of particles, X, across a plane perpendicular to the direction x is given by

$$X = -tD \frac{dn}{dx} \tag{6-2}$$

where t is time, D is the diffusion coefficient, and n is the particle density in the plane. The diffusion coefficient for rigid spherical particles can be calculated from the equation

$$D = \frac{RT(1 + 2A\lambda/d)}{N3\pi\eta d} \tag{6-3}$$

where N is Avogadro's number. The diffusion coefficients of rigid spheres of unit density in air at 20°C and 760 torr for particles having radii of 1.0, 0.4, and 0.1 μm are 2.76×10^{-7}, 8.42×10^{-7}, and 6.84×10^{-6} cm^2 sec^{-1}, respectively. Obviously, the rate of diffusion increases with decreasing particle size, and this principal is the basis of the "diffusion battery" and related instruments, the first of which was developed by Nolan and Guerrini.[7]

Measurements are made of the diffusional deposition of particles from an aerosol undergoing laminar flow through a channel which may be of circular, rectangular, or square cross section. The amount of deposition is usually expressed as the ratio n/n_0 of the particle concentrations at the exit and entrance of the channel, respectively. Various equations have been derived for relating the penetration ratio to the diffusion coefficient and thus to the particle size for monodispersed aerosols. Gormby and Kennedy[8] proposed the following equation for a channel of circular cross section in the early stages of diffusion, that is, when the concentration along the axis is still nearly equal to the initial concentration:

$$\frac{n}{n_0} = 1 - 2.56\ \mu^{2/3} + 1.2\ \mu + 0.177\ \mu^{4/3} \qquad (6\text{-}4)$$

where $\mu = D \times R^2 \bar{v}$, x is the channel length, R is its radius, and $\bar{v}$ is the mean linear flow velocity. The following equation[9] applies for a later stage in the diffusion, when μ exceeds 0.005:

$$\frac{n}{n_0} = 0.819 \exp\ (-3.657\ \mu) + 0.097 \exp\ (-22.3\ \mu) + 0.032 \exp\ (-5.7\ \mu)$$

$$(6\text{-}5)$$

When the channel consists of two parallel plates separated by the distance $2h$, the penetration ratio for both stages is given by the DeMarcus equation:[10]

$$\frac{n}{n_0} = 0.9149 \exp\ (-1.885\ \mu) + 0.0592 \exp\ (-22.3\ \mu)$$

$$+ 0.0258 \exp\ (-151.8\ \mu) \qquad (6\text{-}6)$$

where, for this equation, $\mu = Dx/h^2\bar{v}$. Equations (6-4), (6-5), and (6-6) were derived for monodispersed aerosols. However, most aerosols are polydispersed.

If equations such as (6-5) or (6-6) are used to determine the diffusion coefficient for a polydispersed aerosol, the value obtained will depend on the parameters of the diffusion chamber and on the flow rate, increasing with increasing μ/D. Fuchs and Sutugin[11] have reviewed methods for determining the mean value of the diffusion coefficient of such an aerosol, and the standard deviation of this coefficient. Pollak and Metnieks[12] developed a theory of the diffusion battery method for determining diffusion coefficients (they called it the dynamic method) on the assumption that the diffusion loss of a mixture of particles of different radii is the sum of the diffusion losses of its components. The aerosol is considered to be composed of several monodispersed aerosols, having increasing values of the diffusion coefficients D_1, D_2, $\cdots$, D_i; and the particle concentration fractions p_1, p_2, $\cdots$, p_i. Then, for c parallel rectangular channels,

$$\exp\,(-D'y) = \sum_i p_i \exp\,(-D_iy) \qquad (6\text{-}7)$$

where $y = 1/fQ$, Q is the total volume air flow rate, $f = 3.77^{-1}\,ab^{-1}$ $c^{-1}\,L^{-1}$, $L = $ length, $b = $ height, and $2a = $ width of each channel. When $Q = 0$, $D'/D_s = 1$ where D_s is the smallest value of the D_i. When $Q = \infty$, D' is the weighted average diffusion coefficient of the partial concentrations:

$$D' = \sum_i D_ip_i / \sum_i p_i \qquad (6\text{-}8)$$

Therefore, by measuring the diffusion coefficient for different flow rates, considerable information concerning the mean particle size and the size distribution can be obtained. By determining D_s (in effect subtracting it from the D_i's, redetermining D_s for the aerosol minus the fraction having the smallest diffusion coefficient, and so on) the diffusion coefficients of the "components" in ascending order and their concentrations can be obtained. Fuchs and Sutugin[11] state that the accuracy of the method is not very good because of the need for repeated extrapolations.

Fuchs et al.[13] used a computer to obtain curves for n/n_0 as a function of μ/D from equation (6-6). The size distributions were assumed to be log normal. The "best fit" of one of these curves to the experimental data is made, which yields D and the logarithm of the standard deviation. Several other more-or-less successful methods are described by Fuchs and Sutugin.[11]

Tan and Thomas[14] have reinvestigated the problem of diffusional losses of particles from a fluid flowing through a duct of rectangular cross section. They found small errors in the commonly used equations such as (6-6) and prepared a detailed table showing values of n/n_0 as a function of channel dimensions, diffusion coefficient, and flow rate.

2.2. Instrumentation

Numerous versions of the diffusion battery have been constructed; they commonly consist of vertical, parallel plates. Since such devices are usually applied to particles sufficiently small that diffusion is relatively rapid (<0.1 μm diameter), the loss by sedimentation in such a device is negligible. Pollak and Daly[15] have described such a battery without end pieces or connecting tubing. The penetration fraction n/n_0 is determined with photoelectric condensation nucleus counters.

A commercial instrument is manufactured by Environment One Corp. under the name Diffusion-Denuder. It is actually two devices in one, since it operates in two modes. One is by diffusion and the other is based on the principle that the larger particles in an aerosol have a higher probability of carrying electrical charges than the smaller ones and, therefore, may be removed by their mobility in an electric field. The diffusion and deposition occurs as the aerosol is drawn between two concentric tubes. The ratio n/n_0 is determined with one of the Environment One condensation nucleus counters, which employs light extinction as a measure of the droplet concentration. A graph furnished by the manufacturer of percent transmission versus particle size is used to estimate the mean value of the particle size in the sampled air.

A very simple and effective diffusion battery can be constructed from a cardboard mailing tube. The tube is filled with some type of packing and the ends provided with a convenient inlet and outlet. Glass beads can be used as the packing; Langer, at the National Center for Atmospheric Research, has found rice to be superior to glass beads, presumably because of the different packing characteristics. Of course, such an instrument must be calibrated for various flow rates with aerosols of known particle size.

3. ELECTRICAL PROPERTIES OF THE PARTICLES

Electrical properties of airborne particles are used in various ways to determine their concentrations and size distributions. In the mid-1940s, Guyton[16] built an instrument that involved drawing the aerosol through a glass orifice about 0.8 mm in diameter and allowing the particles to impinge on a metal plate or wire placed in front of the orifice. Individual particles striking the wire produce electrical impulses that are amplified and fire a thyratron tube. The discharges from the thyratron tube are in turn amplified and operate a mechanical counter. Today this would be considered a rather unsophisticated way to process the signals. Guyton observed that the polarity and size of the pulses were independent of the potential of the collector when the particles were solid, but that water droplets gave positive pulses when the collector was at ground potential, no pulses when the wire was at 22 V, and negative pulses when the collector potential exceeded 22 V. It will only count particles having diameters exceeding about 3 μm. Gucker and O'Konski[17] have discussed the theory of this instrument.

The Diffusion-Denuder of the Environment One Corp. was mentioned in the previous section. The denuder mode is based on an instrument designed by Rich,[18] who used the fact that the charge distribution on a population of particles follows Boltzmann's law. The distribution can be represented by the following equation:

$$\frac{N_0}{N_p} = \exp\left(\frac{p^2 e^2}{2rkT}\right) \simeq \exp\left(\frac{2.8p^2 \times 10^{-6}}{r}\right) \tag{6-9}$$

where N_0 is the concentration of uncharged particles, N_p is the concentration of particles of charge p, k is Boltzmann's constant, and e is the electronic charge. When the radius is smaller than 0.05 μm, the uncharged fraction, F_0, of the particles is given by the equation

$$F_0 = \frac{1}{1 + 2(X + X^4 + X^9)} \tag{6-10}$$

where

$$X = \exp\left(\frac{-2.8 \times 10^{-6}}{r}\right) \tag{6-11}$$

For larger radii,

$$F_0 = \frac{1}{1.06(r \times 10^6)^{1/2}} \tag{6-12}$$

The mobility of a particle is defined as the velocity in cm sec^{-1} that a charged particle attains in a field of 1 V cm^{-1}.

$$V = EM \tag{6-13}$$

where E is the field, V is the velocity, and M is the mobility in cm sec^{-1} per volt cm^{-1}. From these equations a graph can be made of N_p/N_0 versus r.

The equivalent particle size of an aerosol is established by determining the total concentration of particles with a condensation nucleus counter. The charged fraction is then determined by passing the aerosol through the electric field of the denuder, and the uncharged fraction, leaving the denuder, is measured. The equivalent size then can be obtained from N_p/N_0 and the graph. The original Rich device makes use of the charges naturally occurring on particles in the air. It consists essentially of two concentric cylinders having a voltage difference of 5–10 kV.

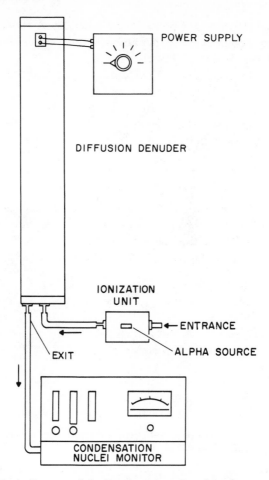

Figure 6-2.   Block diagram of the Environment One denuder.

The Environment One denuder is shown schematically in Figure 6-2. Unlike the original Rich instrument, this unit charges the particles. The aerosol is drawn past a low-level radioactive source and then through the denuder by the pumping action of a condensation nuclei counter. The particles are charged to an equilibrium condition as they pass the radioactive source. The aerosol passes between the same concentric tubes used in the diffusion mode mentioned earlier, but with a voltage applied across the tubes. The condensation nucleus counter at the exit measures the particle concentration, which is compared with the exit concentration when the instrument is operated in the diffusion mode (that is, with no applied voltage). This provides the "percentage transmission" of the larger particles that are not removed by diffusion, and the corresponding particle size is obtained from a graph furnished with the instrument.

An electrostatic cloud droplet sizing device (electrostatic disdrometer) was developed by Keily and Millen,[19] and extensively tested and modified by Abbott et al.[20] The modified version is shown in Figure 6-3. The cloud sample is drawn through the orifice at near sonic velocities by a vacuum pump. Each fog droplet then breaks up into a number of smaller droplets that strike a metal electrode aligned with the axis of the orifice and maintained at a high potential with respect to the surrounding housing. Each group of droplets impacting on the electrode produces a voltage pulse proportional in amplitude to the original droplet size.

The outer shape of the tip of the electrode of the version of Abbott et al. is the forward half of an ellipsoid of revolution with a fineness ratio of 4 and a minor axis of 2.5 cm. Behind the portion shown in Figure 6-3 is a cylindrical section which contains a transistorized preamplifier. This instrument was operated at the National Center for Atmospheric Research on a 75-cm-long boom on the nose of the same glider on which the aerosol camera was mounted (Figure 4-20). The electrical pulses produced by the droplets are amplified and fed into a 10-channel pulse-height analyzer that accumulates and reads out the data each 0.5 sec. The droplet size range that can be measured is about 4–30 μm radius. According to Abbott et al., the principle of operation of the instrument could be readily adapted for measurements in the drizzle and raindrop size range.

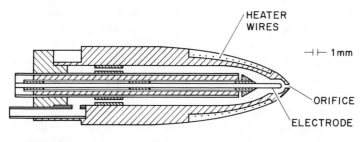

DIA. = 2.5 cm

Figure 6-3. Schematic diagram of the tip of the present model of the electrostatic disdrometer.[20]

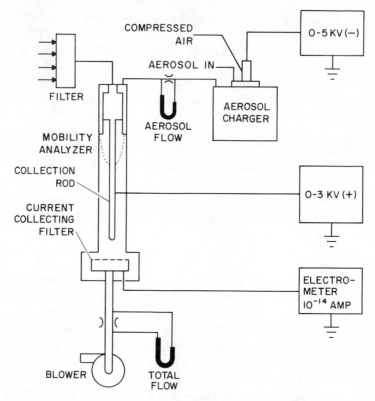

Figure 6-4. Schematic of the Whitby-Clark electrical particle counter system.[21]

Whitby and Clark[21] designed and built an electrical particle counter and size analyzer for the 0.015–1.0 μm diameter range. Figure 6-4 is a schematic drawing of the system. Aerosol particles entering the system are given a unipolar negative charge by a special diffusion charger. The aerosol then passes through a flow meter to a mobility analyzer which it enters as a thin annular cylinder around a core of clean air. The aerosol particles are drawn inward by the positive charge on the collector rod. If the potential on the collector rod is constant, the particles are classified along the rod according to their electrical mobility, but all particles having a mobility less than a given value will miss the rod and will be collected by the current collecting filter. By varying the collector voltage in a systematic manner, the number-particle size distribution curve for the aerosol can be calculated from the filter-current versus collector rod voltage curve. Not all the particles are charged, the percentage decreasing with decreasing particle size. All particles smaller than about 0.05 μm diameter are singly charged. Whitby and Clark used a combination of their counter, a General Electric condensation nucleus counter, and a Royco Model PC 200A optical counter to obtain particle size distributions in the diameter range 0.002–10 μm.

A smaller, automatic version of the Whitby-Clark analyzer has been described by Liu, Whitby, and Pui.[22] The particle size range is 0.0032–1.0 μm diameter and the concentration range is 1–1000 μg m^{-3}, based on typical atmospheric aerosol distributions. It is designed to operate at 0.7–1.2 atmospheres pressure, and in the temperature range -10–50°C. It is manufactured by Thermo-Systems, Inc. (Model 3030).

An electrical method for determining mass concentrations of particles in the atmosphere involves depositing them on a piezoelectric microbalance. Thus mass is measured directly instead of via some parameter of mass.

When a piezoelectric material such as a quartz crystal is electrically driven, it oscillates with a precise natural frequency. This frequency depends on the orientation of the slice of the crystal with respect to the crystal axis, and on its thickness and density. For a given slice, the natural frequency decreases in direct proportion to the mass of any material which may be deposited on it.

Figure 6-5 is an instrument based on this principle designed and built by Olin et al.[23,24] An electrostatic precipitator deposits particles from the aerosol onto the piezoelectric balance. The particles enter along the corona needle, producing a concentric deposit. They are directed through the most intense portion of the corona discharge, producing a high charging efficiency. According to the authors, the mass concentrations of typical atmospheric aerosols can be measured with 5% accuracy in about 10 sec.

Thermo-Systems, Inc. manufactures several versions of this instrument. Their 3200 Particle Mass Monitor Systems include a mass monitor module, a vacuum pump, and either a digital indicator module or a strip chart recorder. These instruments include a reference quartz crystal to compensate for possible changes in ambient conditions, such as temperature and gas composition. The particle size range is 0.01–10 μm in diameter and the particle mass concentration range is 2 to 20,000 μg-m^{-3}. The crystals of such instruments must of course be cleaned after a considerable amount of material has been deposited on them. The maximum sensor loading of the Thermo-Systems instruments is 5–50 μg, depending on the type of particles, and about 78 readings can be made between crystal cleanings.

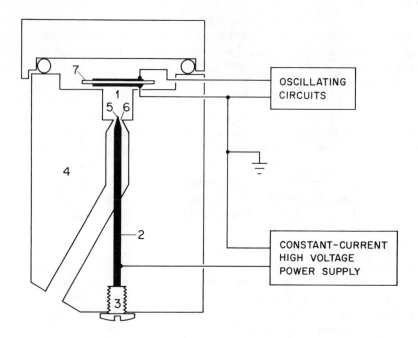

Figure 6-5. Side view of electrostatic precipitator and piezoelectric micro-balance: (1) precipitator chamber, 0.25 inch diameter by 0.25 inch high; (2) corona needle; (3) needle adjusting screw; (4) Teflon precipitating block; (5) high corona region; (6) entrance nozzle to precipitating chamber; (7) piezo-electric crystal sensor.[23] By permission of the American Industrial Hygiene Association.

Celesco Industries, Inc., manufactures a line of these instruments that use electrostatic precipitators or impactors to deposit the particles on the piezoelectric balance. One of their instruments classifies the particles with a multistage impactor having a separate piezoelectric crystal at each stage.

4. ACOUSTIC PARTICLE COUNTERS AND SIZERS

A number of acoustic particle counters and sizers have been developed, but for the most part they have not been very popular. The acoustic

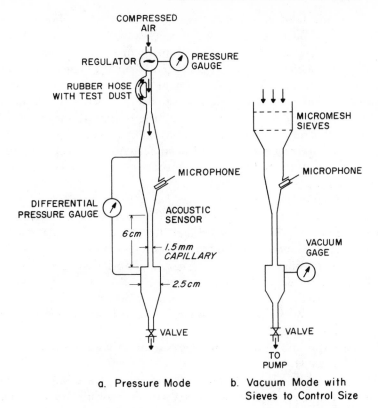

a. Pressure Mode

b. Vacuum Mode with Sieves to Control Size

Figure 6-6.   Langer acoustic counter.[25,26]

counter developed by Langer[25,26] has already been mentioned in connection with his condensation nuclei counter. Two versions of his acoustic counter are shown in Figure 6-6. These devices emit a click as a particle passes through a special glass capillary when the linear flow velocity in the capillary is between 10 and 100 m sec^{-1} and the particle exceeds 30 μm in diameter. The microphone used by Langer was a "subminiature" high-frequency-response device that could be mounted close to the capillary to avoid marked attenuation of the sound. It was purchased from the Thermo Electron Corporation. The frequency response extended to 40,000 cps, and the sensing diaphragm was less than 2 mm in diameter.

Tests with filaments of smoke showed that the flow through the capillary is laminar and breaks up momentarily along a local flow streamline when the sensor is triggered by a particle.

The amplitude of the electrical signal from the microphone was compared with the particle size in the diameter range 10–70 μm. The effect of particle size was within the variation for one size. However, smaller particles gave a shorter and sharper signal than the larger ones. Therefore entire pulse trains from each signal were recorded and examined. The initial fast-rise pulse, which is actually a shock wave, is followed by a second pulse of lower frequency and lesser amplitude, followed by a third pulse, and so on. Langer found that the larger the particles, the larger the amplitude and the lower the frequency of the secondary pulses. However, the problem of quantitatively relating the pulse character to particle size was not solved.

Other research on acoustic counters has been undertaken by Avy and Benarie,[27] who allowed the particles to impact on a microphone diaphragm. This technique has been used for studying interplanetary dust by flying a microphone detector on a rocket. For this purpose the detector has usually consisted of a diaphragm, a piezoelectric crystal, and electronic amplification and storage equipment. A meteoroid striking the diaphragm produces an impulse the magnitude of which is determined by the particle mass, the impact velocity, and the position of the impact on the diaphragm surface.

A theoretical and experimental investigation of Langer's counter was undertaken by Hofman.[28]

Aerosol particle size and concentration can be determined from the attenuation and dispersion of sound.[29] When acoustic waves propagate in an aerosol, they undergo attenuation and dispersion because of the dynamic and thermal lag between the aerosol particles and the gas. When the interactions are entirely viscous and thermal, the attenuation and dispersion coefficients for monodispersed aerosols of small mass concentration are given by the equations

$$\alpha\lambda = 2\pi C_m \left[\frac{\omega\tau_d}{1 + \omega^2\tau_d{}^2} + (\gamma - 1) \frac{C_p{}'}{C_p} \frac{\omega\tau_t}{1 + \omega^2\tau_t{}^2} \right] \qquad (6\text{-}14)$$

and

$$\left(\frac{a_0}{a}\right)^2 - 1 = C_m \left[\frac{1}{1 + \omega^2\tau_d{}^2} + \frac{(\gamma - 1)(C_p{}'/C_p)}{1 + \omega^2\tau_t{}^2} \right] \qquad (6\text{-}15)$$

where λ = wavelength of sound
  a, a_0 = speed of sound in the gas with and without suspended particles
  $C_m = (4/3)\pi r^3 \rho N$ = mass fraction of particles
  $C_p, C_p{}'$ = specific heat of gas at constant pressure, and of particles

α = attenuation coefficient
$\tau_d = (2\rho/9\eta)r^2$ = dynamic relaxation time of particle
η,k = dynamic and kinematic coefficient of viscosity
ρ = particle density
$\tau_t = 3PC_p'\tau_d/2C_p$ = thermal relaxation time of particle
P = Prandtl number of gas = $C_p k/k_a$
k_a = thermal conductivity of the gas
ω = frequency of sound.

Equations (6-14) and (6-15) are valid only when $(\rho'/\rho) \ll 1$ where ρ' is the density of the gas and $(\omega r^2/2k)^{1/2} \ll 1$.

Dobbins and Temkin[29] showed that these equations can be used to obtain particle diameters and mass concentrations by means of various combinations of measurements. One approach is to measure attenuation and dispersion at a single frequency. The ratio of attenuation to dispersion for any given wavelength λ is

$$\gamma = \frac{\alpha\lambda}{(a_0/a)^2 - 1} \tag{6-16}$$

This ratio is a single-valued function of $\omega\tau d$ (6-14 divided by 6-15), and the particle diameter can be calculated from the equation

$$d = (18\eta \, \omega\tau_d/\rho)^{1/2} \tag{6-17}$$

Then the mass fraction C_m can be calculated from either equation (6-14) or (6-15). Similarly, aerosol properties can be calculated from measurements of attenuation at two frequencies or from dispersion at two frequencies.

The measurements of attenuation and dispersion were made with an acoustic interferometer, which consists of a cylindrical tube in which a standing wave pressure field is produced with an acoustic driver.[30] Good agreement was found between the acoustical method and an optical method for particles in the size range about 1–5 μm in diameter, the size range investigated.

Attenuation and dispersion can also be measured by other methods. A pulse technique can be used for large samples of aerosols. This method involves sending pulses through the sample and observing the decrease of pulse amplitude and velocity. Another method suggested by Dobbins and Temkin is that of measuring sound decay in a closed chamber. Since this method furnishes only the coefficient of attenuation, the particle size must be determined by measuring the attenuation for two frequencies.

Another sonic technique for determining the sizes of particles, which can also be used for determining particle size distributions, is based on the principle that particles suspended in air subjected to an intense sound field tend to follow the air molecules and vibrate with the sound wave. Very small particles vibrate with the full amplitude of the wave, larger particles lag behind, and very large particles do not vibrate at all. The ratio of the amplitude of vibration of the particle, x, to that of the sound wave, x_0, is a function of the size of the particle.

The following equation relating x/x_0 to r was derived in connection with studies of sonic coagulation:

$$\frac{x}{x_0} = \left\{ \left[\frac{4\pi r^2 N_1}{9\eta} \left(1 + \frac{A\lambda}{r} \right) \right]^2 + 1 \right\}^{-1/2} \tag{6-18}$$

where λ is now the mean free path of the gas and N_1 is the frequency of the sound.

Instrumentation for determining x/x_0 for individual particles in a suspension has been described by Cassel and Schultz[31] and by Gucker.[32] The former authors used dark-field illumination of the suspended particles, which were photographed through a low-power microscope with a 35mm camera. The sonic field was produced with an 8-inch, 7-W, permanent-magnet loudspeaker activated through an amplifier by an audio-frequency oscillator. The frequencies ranged from 65 to 1500 cps. The photographs revealed oscillating tracks of the particles, and the amplitudes of the oscillations of the smallest particles measured from the track were assumed to be the amplitude of the sound wave, x_0. The amplitudes of the other tracks were the values of x.

Gucker's system was similar. His sonic generator was a modification of the dynamic loudspeaker used by St. Clair.[33]

5. SEDIMENTATION

Sedimentation by gravity has seldom been used for the direct measurement or classification according to size of airborne particles, but there have been a few such applications. One is the Precision Aerosol Spectrometer manufactured by Fleming Instruments, Ltd. (Caston Way, Stevenage, Herts, England). The aerosol is sampled at 0.5, 1, or 2 cm³ min⁻¹ and passes along the axis of a horizontal duct that is part of a closed system through which clean "winnowing" air is circulated at 100 cm³ min⁻¹. The particles are moved along the duct by the winnowing air, settling under gravity. A deposit is produced in decreasing order of size upon a row of glass slides. Spheres of unit density in the size range 1–20 μm in diameter can be separated. The sampling period can be pre-set for periods up to 22 hours. It was designed especially for use in factories and mines.

A much more elaborate, automatic instrument has been described by Samartzopoulos[34] for determining the size distributions of droplets in sprays. The droplet size should be in the range 5–250 μm in diameter. The spray from a nozzle is injected for 0.5–2 sec into the top of a sedimentation tower 6 m high and 0.77 m in diameter. The droplets are collected on the pan of an automatic recording balance at the bottom of the tower.

REFERENCES

1. Cadle, R. D., *Particles in the Atmosphere and Space,* Reinhold, New York, 1966.

2. Pollak, L. W., *Int. J. Air Poll.,* **1**, 293 (1959).

3. Rosen, J. M., Personal communication, 1974.

4. Nolan, P. J., and L. W. Pollak, *Proc. Roy. Irish Acad. Sci.,* **51A2**, 9–31 (1946).

5. Rich, T. A., Geofisica *Pura e Applicata,* **31**, 60 (1955).

6. Langer, G., Personal communication, 1974.

7. Nolan, P., and V. Guerrini, *Proc. Roy. Irish Acad. Sci.,* **43**, 5 (1935).

8. Gormby, P., and M. Kennedy, *Proc. Roy. Irish Acad. Sci.,* **52A**, 166 (1949).

9. Gormby, P., and M. Kennedy, *Proc. Roy. Irish Acad. Sci.,* **45A**, 59 (1938).

10. DeMarcus, W. C., and J. W. Thomas, *Theory of a Diffusion Battery,* U.S. *A.E.C.* Report ORNL-1413 (1952).

11. Fuchs, N. A., and A. G. Sutugin, High-Dispersed Aerosols, in *Topics in Current Aerosol Research,* Vol. 2, G. M. Hidy and J. R. Brock, Eds., Pergamon, New York, 1971.

12. Pollak, L. W., and A. L. Metneiks, *Geofisica Pura e Applicata,* **37**, 183 (1957).

13. Fuchs, N. A., I. Stechkina, and V. Starosselski, *Br. J. Appl. Phys.,* **13**, 281 (1962).

14. Tan, C. W., and J. W. Thomas, *J. Aerosol Sci.,* **3**, 39 (1972).

15. Pollak, L. W., and J. Daly, *Geofisica Pura e Applicata,* **45**, 249 (1960).

16. Guyton, A. C., *J. Ind. Hyg. Toxicol.,* **28**, 133 (1946).

17. Gucker, F. T., Jr., and C. T. O'Konski, *Chem. Revs.,* **44**, 373 (1949).

18. Rich, T. A., *Int. J. Air Poll.,* **1**, 288 (1959).

19. Keily, D. P., and S. G. Millen, *J. Meteor.*, **17**, 349 (1960).

20. Abbott, C. E., J. E. Dye, and J. D. Sartor, *J. Appl. Meteor.*, **11**, 1092 (1972).

21. Whitby, K. T., and W. E. Clark, *Tellus*, **18**, 573 (1966).

22. Liu, B. Y. H., K. T. Whitby, and D. H. U. Pui, presented at the 66th Annual Meeting of the Air Pollution Control Association, Chicago, Ill., June 24–28, 1973.

23. Olin, J. G., G. J. Sem, and D. L. Christenson, *Am. Ind. Hyg. Assoc. J.*, **32**, 209 (1971).

24. Olin, J. G., and G. J. Sem, *Atmos. Environ.*, **5**, 653 (1971).

25. Langer, G., *Powder Technol.*, **2**, 307 (1968/69).

26. Langer, G., *Powder Technol.*, **6**, 5 (1972).

27. Avy, A. P., and M. Benarie, *Staub*, **24**, 343 (1964).

28. Hofman, P., *Staub*, **28**, 360 (1968).

29. Dobbins, R. A., and S. Temkin, *J. Coll. Interface Sci.*, **25**, 329 (1967).

30. Scott, R. A., *Proc. Phys. Soc. (London)*, **58**, 253 (1946).

31. Cassel, H. M., and H. Schultz, in *Air Pollution*, L. McCabe, Ed., McGraw-Hill, New York, 1952.

32. Gucker, F. T., Jr., in *Proceedings of the First National Air Pollution Symposium*, Stanford Research Institute, Stanford, Calif., 1949.

33. St. Clair, H. W., *Rev. Sci. Instr.*, **12**, 250 (1941).

34. Samartzopoulos, C. G., in *Particle Size Analysis*, The Society for Analytical Chemistry, London, 1967.

7

SAMPLING PROBES AND LINES

Most devices for collecting or measuring airborne particles require that a sample of the aerosol be drawn into the device. The sampling system often consists of a probe where the sample is actually withdrawn and some sort of tube or duct to transport the sample from the probe to the analyzer. The sampling system often includes auxiliary equipment such as pitot tubes to measure the flow rate. Such sampling systems must usually be carefully designed and operated in order to obtain representative samples. Criteria for designing such systems are discussed in the following sections.

1. ISOKINETIC SAMPLING

Isokinetic sampling can be defined as sampling by drawing a suspension into a probe at the same linear velocity as that of the bulk of the suspension. Thus both the rate and direction of flow are unchanged by the sampling. Isokinetic sampling decreases errors that result from the failure of larger, heavier particles to follow changes in the direction and velocity of flow of the suspending fluid. Although isokinetic sampling can often be closely approached, it can seldom be strictly achieved. For example, it is nearly impossible to achieve when sampling essentially still air or air that is changing rapidly and erratically in velocity. Also, the walls of any sampling probe disturb the flow of air. Therefore, estimating the errors resulting from various degrees of departure from isokinetic sampling is essential.

319

A number of theoretical and experimental estimations have been made of the errors resulting from anisokinetic sampling. The earliest was probably by Watson,[1] who derived a semiempirical equation:

$$\frac{N}{N_0} = \frac{v_0}{v}\left[1 + f(K)\left\{\left(\frac{v}{v_0}\right)^{1/2} - 1\right\}\right]^2 \qquad (7\text{-}1)$$

where N is the concentration measured, N_0 is the true concentration, v_0 is the stream velocity, v is the mean air velocity at the sampling orifice, K is the inertial parameter, $d^2\rho v_0/18\eta D$, D is the diameter of the orifice, ρ is the specific gravity of the particles, and η is the viscosity of the fluid.

The relationship between $f(K)$ and K was determined experimentally by dispersing lycopodium spores ($d = 32$ μm) and lycoperdon giganteum spores ($d = 4$ μm) in air in a wind tunnel "of low turbulence," and sampling with glass probes. The results are shown in Figures 7-1 and 7-2. Judging from Figure 7-2, the errors from anisokinetic sampling are

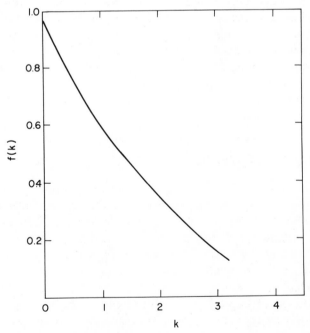

Figure 7.1   Relation between f(K) and K. Reprinted by permission of the American Industrial Hygiene Association.

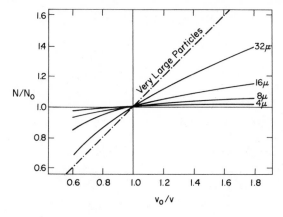

Figure 7-2.   Values of the ratio of observed to actual concentration (N/N_0) for various ratios of air velocity to sampling velocity (v_0/v). The particle diameters which the curves represent are shown on the right. Reprinted by permission of the American Industrial Hygiene Association.

very small for spheres of unit density smaller than 1 or 2 μm in diameter. Equation (7-1) cannot be used when either v_0 or v approach zero.

More recently Badzioch[2,3] has treated the problem of anisokinetic sampling in a semiempirical manner. He started by assuming the validity of the equation

$$\frac{M}{M_0} = \frac{(1 - \alpha)\, v}{v_0 + \alpha} \qquad (7\text{-}2)$$

where M is the mass of particles entering the nozzle at sampling velocity v, and M_0 is the mass of particles entering the nozzle when the sampling is isokinetic. For fine particles, α approaches zero:

$$\frac{M}{M_0} \simeq \frac{v}{v_0} \qquad (7\text{-}3)$$

and for large particles α approaches unity:

$$\frac{M}{M_0} \simeq 1 \qquad (7\text{-}4)$$

Thus for very small or very large particles the errors arising from anisokinetic sampling are small, provided the appropriate method of calculation is used.

In order to estimate α, Badzioch considered a simplified flow pattern leading to the equation

$$\alpha = \left[1 - \exp\left(\frac{-L}{D_s}\right) \right]\left[\frac{L}{D_s} \right]^{-1} \tag{7-5}$$

where L is a length representative of the distance upstream of the nozzle over which there is a disturbance to the gas stream and D_s is the stopping distance, $\rho_p d^2 v_0 / 18\eta$, defined in Chapter II. Experiments showed that L depends only on the diameter of the nozzle.

Badzioch[2] also described procedures for determining the true concentration and size distribution of particles collected anisokinetically.

A somewhat more elaborate theoretical development of the effect of anisokinetic sampling was undertaken by Rüping,[4] who included gravitational settling. He concluded that the importance of isokinetic sampling is, in general, overestimated in the case of dust concentration measurement and also in the case of particle size analysis, other factors being more important.

Davies[5] has investigated the problem of sampling with a probe in calm air, starting with the concept of the relaxation time, τ, defined as $m/6\pi r \eta$ where m is the particle mass. The stopping distance is the initial velocity of the particle in still air times τ. Davies evaluated the inertial effect by defining the stopping distance, D_s, in terms of the air velocity v' at a distance D_s from the center of the orifice:

$$v' = \frac{F}{4\pi D_s{}^2} \tag{7-6}$$

where F is the volume rate of sampling. Then the stopping distance is $v'\tau$, and

$$v' = \left(\frac{F}{4\pi\tau^2}\right)^{1/3} \tag{7-7}$$

For inertial effects to be negligible near the orifice, the stopping distance must be small relative to the radius R of the orifice:

$$\left(\frac{F}{4\pi v'}\right)^{1/2} \ll R \tag{7-8}$$

and

$$\left(\frac{F\tau}{2\pi}\right)^{1/3} \ll R \tag{7-9}$$

The condition for the orientation of the orifice not to matter is

$$\frac{F}{\pi R^2} \gg g\tau \qquad (7\text{-}10)$$

where g is the acceleration due to gravity, or

$$R \ll \left(\frac{F}{\pi g \tau}\right)^{1/2} \qquad (7\text{-}11)$$

Equations (7-10) and (7-11) are derived from Stokes' equation for the terminal velocity of falling particles.

Thus the conditions for obtaining a true sample when sampling through a small tube in still air are

$$\left(\frac{F\tau}{4\pi}\right)^{1/3} \ll R \ll \left(\frac{F}{\pi g \tau}\right)^{1/2} \qquad (7\text{-}12)$$

Davies constructed Table 7-1 using the limiting criteria that the inertial factor on the left is $R/5$ and the sedimentation factor on the right is $5R$, that is,

$$5\left(\frac{F\tau}{4\pi}\right)^{1/3} \leq R \leq \frac{1}{5}\left(\frac{F}{\pi g \tau}\right)^{1/3}$$

Using these limiting criteria, the greatest errors due to sedimentation, occurring when the tube orifice points directly upward or downward, are, respectively, a 4% overestimate or underestimate of particle concentration resulting from sedimentation with or against the flow into the orifice. Davies estimated that within these limits the error due to inertial effects is less than 1.6%.

Davies also estimated the errors occurring when sampling with a small tube in a wind. Since a steady wind, blowing horizontally, imposes a horizontal drift upon aerosol particles, whereas gravity makes them descend vertically, these motions can be considered independently and simply added to the motion due to suction into the orifice as long as the inertia of the particles has only a small effect. Davies concluded that the conditions of Table 7-1 apply when the wind velocity is less than $v'/5$.

2. SAMPLING LINES

Once a sample of aerosol has entered a sampling probe and is being drawn along a tube or duct leading the sample to a particle-collecting

Table 7-1. Permissible Radii* of Tubes (cm) for Sampling Aerosols in Calm Conditions[5]

Particle diameter, μm	Rate of suction, F (cm³/sec)					
	1	10	100	1,000	10,000	100,000
1	0.033 – 1.9	0.071 – 6.0	0.15 – 19	0.33 – 60	0.71 – 190	1.5 – 600
2	0.051 – 1.0	0.11 – 3.2	0.23 – 10	0.51 – 32	1.1 – 100	2.3 – 320
5	0.093 – 0.41	0.20 – 1.3	0.43 – 4.1	0.93 – 13	2.0 – 41	4.3 – 130
10	0.15 – 0.21	0.31 – 0.65	0.68 – 2.1	1.5 – 6.5	3.1 – 21	6.8 – 65
20	(0.23 ~ 0.10)	(0.50 ~ 0.33)	(1.1 ~ 1.0)	2.3 – 3.1	5.0 – 10.3	11.0 – 31
50	(0.42 ~ 0.042)	(0.90 ~ 0.13)	(1.9 ~ 0.42)	(4.2 ~ 1.33)	(9.0 ~ 4.2)	(19 ~ 13.3)
100	(0.63 ~ 0.023)	(1.4 ~ 0.071)	(2.9 ~ 0.23)	(6.3 ~ 0.71)	(14.0 ~ 2.3)	(29 ~ 7.1)
200	(0.89 ~ 0.014)	(1.9 ~ 0.037)	(4.1 ~ 0.14)	(8.9 ~ 0.37)	(19 ~ 1.4)	(41 ~ 3.7)
500	(1.26 ~ 0.008)	(2.7 ~ 0.025)	(5.8 ~ 0.08)	(12.6 ~ 0.25)	(27 ~ 0.80)	(58 ~ 2.5)

* When the radius is greater than the left hand figure, error due to particle inertia is negligible; when the radius is smaller than the right hand one, the settlement of particles has a negligible effect. The two criteria can be satisfied simultaneously for the entries in the upper part of the table, but this is not possible for the entries in brackets. Satisfactory samples can be obtained for the latter by selecting a tube with a radius exceeding the left hand value and using it so that the plane of the orifice is vertical. The entries in the upper part of the table constitute small tube sampling.

device or measuring instrument, numerous mechanisms can operate to remove particles from the aerosol, producing erroneous results. These mechanisms include sedimentation of large particles, impaction on the walls at bends, and diffusion of small particles to the walls. These mechanisms, and the theory which can be used to estimate the importance of these mechanisms for a particular sampling system and aerosol, are discussed in the preceding chapters. In general, ducts should be as short as possible and for some purposes their use can be avoided entirely, for example when sampling with open-face filter holders. Sharp bends should be avoided to decrease impaction (centrifugal separation) at the bends, and laminar flow through the tube should be maintained. Usually large tubes or ducts are preferable to small ones (but refer to Table 7-1 with regard to the tube orifice).

3. INSTRUMENTATION

A great variety of sampling probes has been developed for various types of aerosol sampling, and even for the same type of sampling such as that from stacks. Such devices are often referred to as isokinetic probes or isokinetic samplers. Any attempt to describe and comment on more than a small fraction of these would be hopeless and repetitious, but a few examples can be described.

A sampling system that includes isokinetic sampling and collection of "source," (that is, stack, etc. emissions, both particulate and gaseous) has been developed by the U.S. Environmental Protection Agency.[6] Modified versions are manufactured by the Western Precipitation Division of the Joy Manufacturing Co. and by Lear Siegler, Inc. The basic unit consists of a combined probe and pitot tube (the pitobe), a sampling unit, a control unit, a rail assembly for supporting the probe, and an "umbilical" cord for remote control. Particle collection is by a filter assembly and an optional cyclone assembly. Figure 7-3 is a diagram of the Western Precipitation system. Other optional equipment includes a meter rate computer, a portable refrigeration system, and a condenser. The umbilical cord contains all electrical interconnections plus a vacuum line and two smaller lines for pitobe pressure measurements. The condenser assembly removes moisture and other condensibles from humid, hot gas streams, protecting pumps and dry gas meters from contamination and corrosion.

Boubel[7] has designed a source particle sampling train designed to fill the need for a sampler that can obtain a reliable sample from a variety

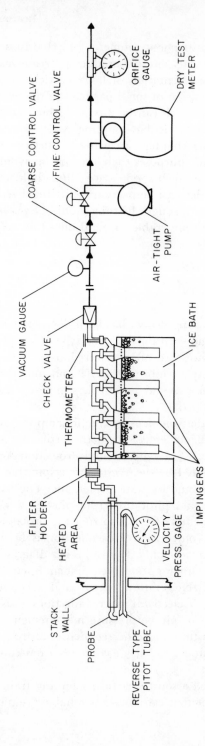

Figure 7.3 Western Precipitation isokinetic sampling train.

326

of sources cheaply, quickly, and under field conditions. A high-volume filter sampler was used to collect the particles since Government emission standards are based on a mathematical model related to the desired ambient air standards, and it seemed desirable to employ the same type of sampler to obtain both the atmospheric particulate loading and the emission particulate loading from the sources in the same region.

Boubel's sampler, shown in Figure 7-4, is constructed of aluminum and is sufficiently light that it can be held by one person during sampling. Like the EPA system, it includes an integral pitot tube, which is connected to a gage for simultaneous velocity measurement and sample collection. Another similarity to the EPA system is that the suction unit is separated from the probe by a length of flexible hose to lighten the assembly. A commercial version is marketed by the Air Sampling Division of Rader Pneumatics, Portland, Oregon.

The following standard procedure was recommended by Boubel. Before use, residual particulate material is removed by washing with methyl alcohol. Then the filters to be used plus a blank for every four filters are conditioned at room temperature in a desiccator for at least twelve hours and weighed. A convenient procedure which prevents loss of the sample is to fold the 8×10 in. glass fiber filters once, collection side in, and place them in 6.5×9.5 in. manila envelopes. The envelopes and filters are then weighed together.

The first step in actual sampling involves placing a temperature-measuring instrument and an Orsat sampling line into the stack. The velocity of the stack gases is then measured using the pitot tube, and a sampling location is chosen such that the velocity is within the design range of the probe. A preliminary sample is then taken at approximately isokinetic conditions to determine the average temperature through the

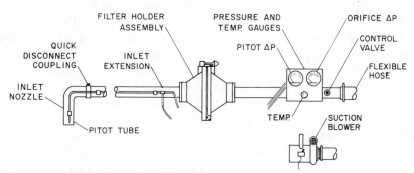

Figure 7-4.   Boubel[7] particle sampler.

filter and orifice. These values are used to obtain isokinetic sampling rates by means of a set of operating curves. The Orsat sampler is used to obtain a gas analysis if that is desired. After sampling, the filters are returned to the laboratory where they are again conditioned at the same temperature and humidity as employed for the original conditioning, and weighed. The probe is rinsed with methyl alcohol, the alcohol is evaporated, and the weight of the residue is divided on a time-weighted basis among the filters used. The emission rates of particulate material from the stack are then calculated from the weight gains of the filters, corrected for any blank filter weight change, plus the material washed from the probe.

A predecessor of such samplers was the system developed by Lapple.[8] It made use of a two-liquid "null point" differential manometer to obtain automatic velocity balancing. The aerosol flow rate through the sampling line was adjusted to maintain the interface in the manometer at a zero point, which represented equal sampling and stack velocities. Particles were collected by filters consisting of porous paper bags inserted in the line between the probe and a metering orifice. Lapple recommended that when the stack gases are too hot to permit the use of the paper bags, the particles be collected with a small cyclone separator followed by water scrubbers.

To obtain a representative sampling of stack gases, samples should be taken at several points across a duct or stack. One approach is to divide the cross-sectional area into a number of imaginary units of equal area. Sampling is carried out at the center of each unit for equal time periods.

Numerous isokinetic samplers have been designed for aircraft. One such instrument has been described by Goodale, Corda, and Evans.[9] The aerosol particles were collected by electrostatic precipitation. The sampler was designed to be mounted outside of and at the front of the aircraft so that it would be in a region of laminar flow instead of in the turbulent boundary region which surrounds the aircraft. The inlet was sharp edged, mounted on gimbal rings, and vaned to ensure its alignment with the direction of air flow. The sample stream was led from the vaned inlet to an expansion section which consisted of a smooth cone with a divergent angle of 7°. This section reduced the sample velocity from 150 to about 3.7 mph, at which velocity the electrostatic precipitator efficiently collected the particles. The particle-free air reentered the atmosphere through an annular outlet around a choke which served as a valve to control the flow rate through the instrument. The external airstream flowing by this outlet provided sufficient suction to compensate for internal friction and ensure isokinetic flow at the inlet.

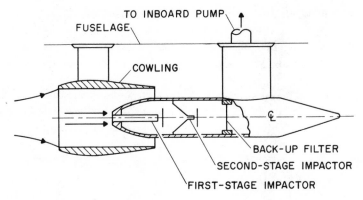

Figure 7-5.   Schematic diagram of isokinetic impactor probe.[10]

Torgeson and Stern[10] designed and built the isokinetic sampler and impactor for aircraft shown schematically in Figure 7-5. This work was an outgrowth of earlier work by Stern, Zeller, and Schekman[11] demonstrating the decrease in the "cut-off" size for particles collected by an impactor when the inertial parameter is increased by decreasing the pressure, as mentioned in Chapter 2. Three particle collection stages were employed, two of them impactors and the third a filter.

The nozzle velocity required for the first-stage impactor was less than one-half the flight speed of the aircraft; thus to achieve isokinetic sampling, the flow was decelerated ahead of the sampler inlet with the cowling shown in the figure. Because of the gradual deceleration of the inlet airstream ahead of the cowling the streamlines near the axis were nearly straight. The use of the cowling also decreased departures from isokinetic sampling due to changes in the angle of attack.

REFERENCES

1. Watson, H. H., *Am. Ind. Hyg. Assoc. Q.,* **15,** 21 (1954).
2. Badzioch, S., *J. Instit. Fuel,* **33,** 106 (1960).
3. Badzioch, S., *Brit. J. Appl. Phys.,* **10,** 26 (1959).
4. Rüping, G., *Staub,* **28,** 137 (1968).
5. Davies, C. N., *Staub,* **28,** 219 (1968).
6. Martin, R. M., Construction details of isokinetic source-sampling equipment, Environmental Protection Agency, Research Triangle Park, North Carolina, April, 1971.

7. Boubel, R. W., *J. Air Poll. Control Assoc.,* **21,** 783 (1971).

8. Lapple, C. E., *Heating, Piping, Air Conditioning,* (July, August, October, November, 1944; December, 1945; February, 1946).

9. Goodale, T. C., B. M. Corda, and E. C. Evans, *Am. Ind. Hyg. Assoc. Q.,* **13,** 226 (1952).

10. Torgeson, W. L., and S. C. Stern, *J. Appl. Meteor,* **5,** 205 (1966).

11. Stern, S. C., H. W. Zeller, and A. I. Schekman, *Ind. Eng. Chem. Funda.,* **1,** 273 (1962).

AUTHOR INDEX

331

Grey, D. S., 170, (232)
Gruber, C. W., 228, (233)
Gucker, F. T., Jr., 238, 239, 240, 250, (277) (278), 316, (318)
Guerrini, V., 303, (317)
Gumprecht, R. O., 272, (279)
Guyton, A. C., 306, (317)

Haines, G. F., Jr., 228, (233)
Hald, A., 25
Hale, W. E., 229, (233)
Hall, C. E., 212, 213, 214, (233)
Hall, J. S., 266, (279)
Hamilton, R. J., 187, (232)
Hansen, J. E., 267, (279)
Harris, G. W., 97, (136)
Hausner, H. H., 8, (43)
Hawkes, P. W., 206, 218, (233)
Heller, W., 238, (277)
Hemeon, W. L. C., 228, (233)
Herdan, G., 9, 12, 18, 21, (43)
Herrmann, K., 52, 53, (133)
Hertel, K. L., 60, (134)
Hewitt, G. W., 126, (137)
Heywood, H., 7, 10, (43)
Hidy, G. M., 40, 41, (44)
Hill, K. C., 83, (135)
Hiltner, W. A., 266, (279)
Hochreiner, D., 295, 296, 297, (299)
Hodkinson, J. R., 237, 242, 254, 270, 271, 272, (277) (278)
Hodsworth, J. F., 187, (232)
Hoel, P. G., 15, (44)
Hofman, P., 314, (318)
Holden, F. R., 200, (232)
Horvath, H., 248, 249, (277)
Hovenier, J. V., 267, (279)
Husar, R. B., 249, (277)

Ide, H. M., 228, (233)
Intelmann, W., 92, (136)
Irani, R. R., 1, 12, (43), 139, 194, (231) (232)

Jacobs, M. B., 81, 82, (135)
Jesse, A., 196, 197, (233)
Johnson, G. G., Jr., 220, (233)
Johnstone, H. F., 53, 59, 94, (134) (136)
Jordon, B. W., 104, 105, (136)
Junge, C. E., 32, 36, 37, 38, (44), 107, 109,
110, (136)
Jupnik, J., 164, (232)

Kallai, T., 293, 294, 295, (299)
Kallman, H., 126, (137)
Kalmus, E. H., 211, (233)
Kanter, C. V., 95, 96, (136)
Katz, S., 250, (278)
Keafer, D., 256, (278)
Keily, D. P., 309, (318)
Kellogg, W. W., 70, 73, 80, 83, (135), 272, 273, (279)
Kennedy, M., 304, (317)
Kerker, M., 27, (44), 235, 237, 239, 241, 242, 244, 269, 270, (276) (277) (279)
Kitani, S., 269, (279)
Knollenberg, R. G., 276, (280)
Knudsen, M., 77, (135)
Koechner, W., 263, (278)
Kondratyev, K. Y., 272, (279)
Kratokvil, J. P., 270, (279)
Krinsley, D. H., 216, (233)
Kurfis, K. R., 272, (279)
Kynaston, D., 205, (233)

Lamb, H., 51, (133)
LaMer, V. K., 102, 118, 120, (136), 243, 244, 250, (277) (278)
Landahl, H., 52, 53, (133)
Lane, B. W., 82, (135)
Lane, W. R., 36, (44), 79, 84, 92, (135), 292, 293, (298)
Langer, G., 303, 313, (317) (318)
Langmuir, I., 52, 54, 55, 56, 57, 58, (134)
Lapple, C. E., 200, (232), 328, (330)
Lazrus, A. L., 83, (135)
Lee, P. K., 250, (278)
Levit, A. B., 28, 29, 30, (44)
Lieberman, A., 250, (278)
Lilienfeld, P., 128, 129, (137)
Linke, F., 273, (279)
Linsky, B., 227, (233)
Lippmann, M., 126, 128, 129, (137)
Lippmann, S. M., 284, (298)
Littlefield, J. B., 114, (136)
Liu, B. Y. H., 102, 132, (136) (137), 249, 255, 256, (277) (278)
Lodge, J. P., Jr., 3, (43), 62, 63, 64, 66, 68, 69, 73, 74, 83, (134) (135)
Lorange, E., 83, (135)

SUBJECT INDEX